# The Migrati... ...e
# 2
# Prog

Organised with support of:

# The Migration Conference 2019 Programme

compiled by
The Migration Conference Team

TRANSNATIONAL PRESS LONDON
2019

The Migration Conference 2019 Programme

Compiled by The Migration Conference Committee

Copyright © 2019 by Transnational Press London

All rights reserved.

First Published in 2019 by TRANSNATIONAL PRESS LONDON in the United Kingdom, 12 Ridgeway Gardens, London, N6 5XR, UK.

www.tplondon.com

Transnational Press London® and the logo and its affiliated brands are registered trademarks.

This book or any portion thereof may not be reproduced or used in any manner whatsoever without the express written permission of the publisher except for the use of brief quotations in a book review or scholarly journal.

Requests for permission to reproduce material from this work should be sent to: sales@tplondon.com

Paperback

ISBN: 978-1-912997-19-0

Cover Design: Gizem Çakır

www.tplondon.com

Under the Patronage of International Organization for Migration

Under the Patronage of President of Puglia Regional Government

Under the Patronage of the Municipality of Bari

## Chair's Welcome

Dear colleagues,

We're pleased to welcome you to the Department of Political Science at the University of Bari "Aldo Moro" for the 7th Migration Conference. The conference is the largest scholarly gathering on migration with a global scope. Human mobility, economics, work, employment, integration, insecurity, diversity and minorities, as well as spatial patterns, culture, arts and legal and political aspects appear to be key areas in the current migration debates and research.

Throughout the program of the Migration Conference you will find various key thematic areas covered in 598 presentations by 767 contributors coming from all around the world, from Australia to Canada, China to Colombia, Brazil to Korea, and South Africa to Norway. We are proud to bring together experts from universities, independent research organisations, governments, NGOs and the media.

We are also proud to bring you opportunities to meet with some of the leading scholars in the field. This year invited speakers include Fiona B. **Adamson**, Markus **Kotzur**, Philip L. **Martin**, Karsten **Paerregaard**, Ferruccio **Pastore**, Martin **Ruhs**, Jeffrey H. **Cohen**, and Carlos Vargas **Silva**.

Although the main language of the conference is English, this year we will have linguistic diversity as usual and there will be presentations in French, Italian, Spanish and Turkish.

We have maintained over the years a frank and friendly environment where constructive criticism foster scholarship, while being nice improves networks and quality of the event. We hope to continue with this tradition and you will enjoy the Conference and Bari during your stay.

We thank all participants, invited speakers and conference committees for their efforts and contribution. We also thank many colleagues who were interested in and submitted abstracts but could not make it this year. We are particularly grateful to hundreds of colleagues who served as reviewers and helped the selection process.

We also thank to those colleagues who organised panels and agreed to chair parallel sessions over three days. We reserve our final thanks to the team of volunteers whose contributions have been essential to the success of the conference. In this regard, special thanks are reserved for our volunteers and team leaders Rosa, Alda, Franco, and Aldo from the University of Bari, Tuncay and Fatma from Regent's University London, Fethiye from Namik Kemal University and Vildan from Galatasaray University, Ege from Middle East Technical University, Mehari from Regent's University London, and Gizem from Transnational Press London.

Our final thanks are reserved for the leaders of the University of Bari "Aldo Moro" and the Department of Political Science, President of Puglia Regional Administration and Mayor of City of Bari for hosting the Conference and for their generous support in enriching the Conference programme.

Please do not hesitate to get in touch with us through the conference email (migrationscholar@gmail.com).

Ibrahim **Sirkeci** and Michela C. **Pellicani**

*The Migration Conference Chairs*


## The Migration Conference 2019

The Migration Conference is a global venue for academics, policy makers, practitioners, students and everybody who is interested in intelligent debate and research informed discussions on human mobility and its impacts around the world. The Migration Conference 2019 is the 7th conference in the series and co-organised and hosted by the University of Bari "Aldo Moro", Italy and Transnational Press London. The Migration Conferences were launched at the Regent's Centre for Transnational Studies in 2012 when the first large scale well attended international peer-reviewed conference with a focus on Turkish migration in Europe in Regent's Park campus of Regent's University London. The migration conferences have been attended by thousands of participants coming from all around the world in London (2012), London (2014), Prague (2015), Vienna (2016), Athens (2017), Lisbon (2018), and Bari (2019).

The Migration Conference 2019 is organised with funding raised by registration fees.

The following organisations have supported the TMC 2019:

Department of Political Science and the University of Bari "Aldo Moro", Italy, Puglia Regional Government, City of Bari, Italian Institute of Statistics, Red Cross Bari Committe, International Organisation for Migration, Italy, Caritas Bari Bitonto, Mediterranean Universities Union, Association Marocaine d'Etudes & de Recherches sur les Migrations, Morocco, Ordine degli Avvocati di Bari, Ordine Assistenti Sociali Regione Puglia, Italy, Global Migration Project at Ohio State University, USA, Regent's University London Centre for Transnational Business and Management, UK, Albrecht Mendelssohn Bartholdy Graduate School of Law, Germany, Institut de Recherche, Formation et Action sur les Migrations, Belgium, Migration Institute of Finland, Transnational Press London, Migration Letters, Remittances Review, Border Crossing, Kurdish Studies, Journal of Gypsy Studies, and Göç Dergisi journals.

The abbreviated title of The Migration Conference is: TMC2019

Hash tag #tmc19bari

 migrationconference.net        tplondon.com

 @migrationevent        fb.me/MigrationConference

Email: migrationscholar@gmail.com

# CONTENT and TIMETABLE[*]

---

[*] [D] indicates developmental papers and sessions with developmental papers; [LANGUAGE] indicates if the session is run in Italian, Spanish or Turkish. All other sessions are in English; [P] indicates poster sessions.

# People
## Conference Executive Committee
Prof Michela C. Pellicani, University of Bari "Aldo Moro", Italy (Chair)
Prof Ibrahim Sirkeci, Regent's University London, UK (Chair)
Prof Jeffrey H. Cohen, Ohio State University, USA
Prof Philip L Martin, University of California Davis, USA
Prof Ali Tilbe, Namik Kemal University, Turkey

## Transnational Advisory Committee
Prof Deborah Anker, Harvard University, United States
Prof Petra Bendel, Friedrich-Alexander University of Erlangen-Nuremberg, Germany
Prof Ali Caglar, Hacettepe University, Turkey
Prof Giovanni Cellamare, University of Bari "Aldo Moro", Italy
Prof Dilek Cindoglu, Hamad Bin Khalifa University, Qatar
Prof Aron Anselem Cohen, University of Granada, Spain
Prof Lucinda Fonseca, University of Lisbon, Portugal
Prof Elli Heikkila, Migration Institute of Finland, Finland
Prof Monica Ibáñez-Angulo, University of Burgos, Spain
Prof Liliana Jubilut, Universidade Católica de Santos, Brazil
Prof Mohamed Khachani, AMERM & University of Rabat, Morocco
Prof Beatrice Knerr, University of Kassel, Germany
Prof Byron Kotzamanis, University of Thessaly, Greece
Prof Jonathan Liu, Regent's University London, UK
Prof Apostolos G Papadopoulos, Harokopio University of Athens, Greece
Prof João Peixoto, University of Lisbon, Portugal
Prof Daniele Petrosino, University of Bari "Aldo Moro", Italy
Prof Nicholas Procter, University of South Australia, Australia
Prof Giuseppe Sciortino, University of Trento, Italy
Prof Claude Sumata, National Pedagogical University, DR Congo

## Transnational Scientific Committee
Dr Nirmala Devi Arunasalam, University of Plymouth, United Kingdom
Dr Bahar Baser, Coventry University, United Kingdom
Dr Bharati Basu, Central Michigan University, United States
Dr Gülseli Baysu, Queen's University Belfast, United Kingdom
Dr Gul Ince Beqo, Universita Cattolica del Sacro Cuore di Milano, Italy
Dr Tuncay Bilecen, Kocaeli University, Turkey & Regent's University London, UK
Dr Elias Boukrami, Regent's University London, United Kingdom
Dr Yaprak Civelek, Istanbul Arel University, Turkey
Dr Nicola Daniele Coniglio, University of Bari "Aldo Moro", Italy
Dr Martina Cvajner, University of Trento, Italy
Dr Carla de Tona, Independent Researcher, Italy
Dr Saniye Dedeoğlu, Mugla Sitki Kocman University, Turkey
Dr Sureya Sonmez Efe, University of Lincoln, United Kingdom
Dr Tahire Erman, Bilkent University, Turkey
Dr Alina Esteves, Universidade de Lisboa, Portugal
Dr Ana Vila Freyer, Universidad Latina de México, Mexico
Dr Gökçe Bayindir Goularas, Yeditepe University, Turkey
Dr Olga R. Gulina, RUSMPI- Institute on Migration Policy, Russia
Dr Sarah E. Hackett, Bath Spa University, United Kingdom
Dr Serena Hussain, Coventry University, United Kingdom
Dr İnci Aksu Kargın, Uşak University, Turkey
Dr Rania Rafik Khalil, The British University in Egypt, Egypt

Dr Emre Eren Korkmaz, University of Oxford, United Kingdom
Dr Alberto Capote Lama, University of Granada, Spain
Dr Simeon Magliveras, King Fahd University of Petroleum and Minerals, Saudi Arabia
Dr Vildan Mahmutoğlu, Galatasaray University, Turkey
Dr Altay Manço, Institut de Recherche, Formation et Action sur les Migrations, Belgium
Dr Marius Matichescu, West Timisoara University, Romania
Dr M. Gökay Özerim, Yaşar University, Turkey
Dr Marjan Petreski, American College Skopje, Macedonia
Isabella Piracci, Avvocatura Generale dello Stato, Rome, Italy
Dr Md Mizanur Rahman, Qatar University, Qatar
Dr Bradley Saunders, Prince Mohammad bin Fahd University, Saudi Arabia
Dr Paulette K. Schuster, Hebrew University Jerusalem, Israel
Dr Deniz Ş. Sert, Özyeğin University, Turkey
Dr Ruchi Singh, Prin. L. N. Welingkar Institute of Management Dev. & Res., India
Dr Armagan Teke Lloyd, Abdullah Gul University, Turkey
Dr Fethiye Tilbe, Namik Kemal University, Turkey
Dr AKM Ahsan Ullah, University Brunei Darussalam, Brunei Darussalam
Dr K. Onur Unutulmaz, Social Sciences University of Ankara, Turkey
Dr Deniz Eroglu Utku, Trakya University
Dr Raffaele Vacca, University of Florida, USA
Dr Hassan Vatanparast, Saskatchewan University, Canada
Dr Pınar Yazgan, Sakarya University, Turkey
Dr M. Murat Yüceşahin, Ankara University, Turkey
Dr Welat Zeydanlioglu, Sweden
Dr Ayman Zohry, Egyptian Society for Migration Studies, Egypt
Emilia Lana de Freitas Castro, Universität Hamburg, Germany
Ülkü Sezgi Sözen, Albrecht Mendelssohn Bartholdy Graduate School of Law, Germany

**Local Organisation Committee**
Michela C. Pellicani, University of Bari "Aldo Moro", Italy (Chair)
Alda Kushi, University of Bari "Aldo Moro", Italy (Logistics contact)
Rosa Venisti, University of Bari "Aldo Moro", Italy (Logistics contact)
Ege Cakir, Middle East Technical University, Turkey (Registration desk)
Mehari Fisseha, Regent's University London, UK (Registration desk)
Irene Albamonte, University of Bari "Aldo Moro", Italy
Gianfranco Berardi, University of Bari "Aldo Moro", Italy
Vittorio Bisceglie, University of Bari "Aldo Moro", Italy
Raffaella Bonerba, University of Bari "Aldo Moro", Italy
Rosanna Bray, University of Bari "Aldo Moro", Italy
Francesco Carlucci, University of Bari "Aldo Moro", Italy
Orsola Castoro, University of Bari "Aldo Moro", Italy
Mario Colonna, University of Bari "Aldo Moro", Italy
Marco Di Sapia, University of Bari "Aldo Moro", Italy
Francesca Falsetti, University of Bari "Aldo Moro", Italy
Ricardo Leonetti, University of Bari "Aldo Moro", Italy
Giandomenico Liano, University of Bari "Aldo Moro", Italy
Concetta Masellis, University of Bari "Aldo Moro", Italy
Aldo Perri, University of Bari "Aldo Moro", Italy
Giacomo Signorile, University of Bari "Aldo Moro", Italy
Francesco Silecchia, University of Bari "Aldo Moro", Italy
Sandro Spataro, University of Bari "Aldo Moro", Italy

## Venue: Department of Political Science, University of Bari

The TMC 2019 Conference venue is the Department of Political Science, University of Bari "Aldo Moro" in Bari, Italy.

Bari is Puglia's regional capital in South Italy. Bari and the surrounding areas are full of touristic attractions to be explored. For details please visit the website of Tourism Office of Puglia: https://www.agenziapugliapromozione.it/portal/en/home.

**The Venue Address**: Dipartimento di Scienze Politiche, Università degli Studi di Bari "Aldo Moro", Piazza Cesare Battisti, 3, 70121 Bari BA, Italy.

The registration desk area, parallel session rooms and plenary session halls for the TMC2019 can be found on the ground floor of the Department of Political Science. Please see the maps below for details.

### Useful information

Emergency phone numbers to contact

Carabinieri – Pronto intervento: 112

Police (Polizia di Stato): 113

Ambulance: 118

Fire Brigade: 115

Airoport: 0039.080.5800200 (same number for all airports of the Region)

Border Police: 0039.080.5316196

Hospital (Policlinico): 0039.080.5575724

Emergency phone numbers work 24 hours per day, 7 days a week, every day.

They are free of charge and they don't need any international code.

### Opening hours

Public administration: Monday to Friday 8.30 – 15.30 or 17.30.

Banks: Monday to Friday 8.30 – 15.30 or 17.30. Banks are closed on Saturday and Sunday.

Pharmacies: Monday to Friday 8.30 – 13.00 and 17.00 – 20.00. Some are open also on Saturday and Sunday with no restriction of time.

### Travel Agency

If you wish to organise a trip in Apulia Region or elsewhere, you can contact:

Travel Agency  Systemar Viaggi

Via Andrea da Bari n. 129

70121  Bari – Italy

Telephone  +39.080.5234330 +39.080.5232574

e-mail: monica@systemarviaggi.it

## Bari City Map:

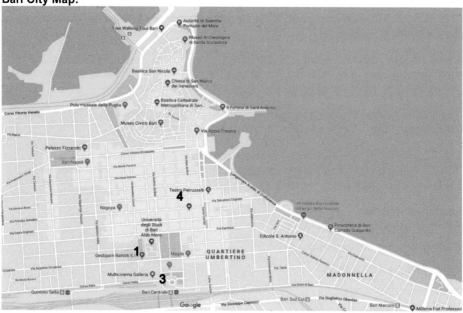

## University of Bari Location:

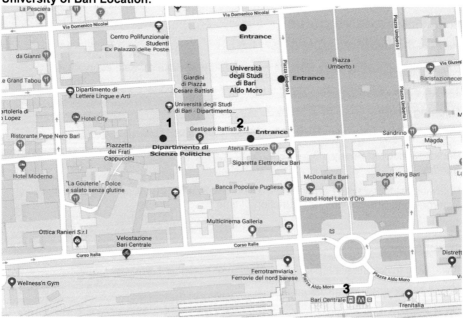

## Guide for the maps:

1 Palazzo Del Prete, Department of Political Science – For parallel sessions
2 Palazzo Ateneo – For Welcoming Speeches, Exhibition, Concert
3 Bari Central Train Station
4 Teatro Petruzelli – for Gala Dinner

**Ground Floor, Department of Political Science:**

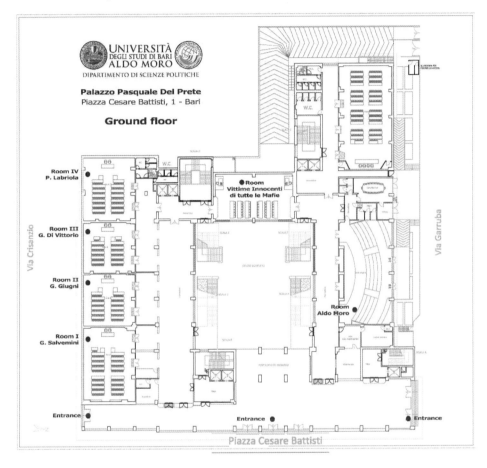

## Second Floor, Department of Political Science:

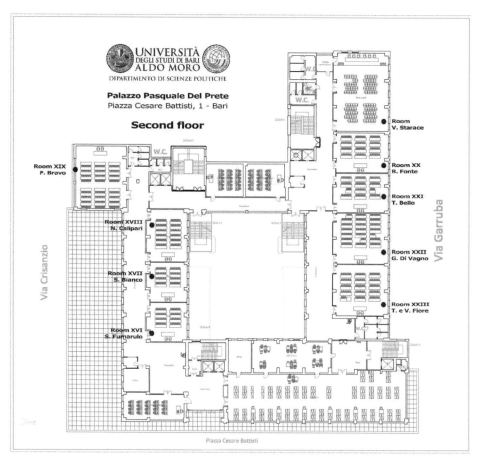

## Conference Gala Dinner:

The Gala Dinner is scheduled for 20th June 2019 from 20:00 to 22:30 at **Circolo Unione - Teatro Petruzzelli**, Corso Cavour, 12, 70122 Bari, Italy.

It is about 10 minutes walking distance from the conference venue.

Tel: + 39 (080) 521 1249

https://www.facebook.com/pages/category/Arts---Entertainment/Circolo-Unione-Teatro-Petruzzelli-392710240812854/

## Keynote speakers at TMC 2019

The Migration Conferences team are proud to have leading scholars in the field as keynote speakers whose details are listed below.

**Markus Kotzur**

### We, the People on the Move – the Impact of International Human Rights Guarantees on Migration Governance

The need for managing migratory movements is evident. However, as desirable a reform of international law instruments, a new international organisation for migration and refugee questions (or at least binding treaty), the inclusion of an explicit right to asylum in human rights treaties, the taking up of climate refugees in the 1951 Convention, and, in particular, adequate normative responses to mass refugee phenomena would be, it seems rather unlikely that the international community is prepared to dare such far-reaching steps. Thus, a consistent "piecemeal engineering" in re migration governance will be the more promising alternative. The paper argues for activation existing human rights as effective instrument for a step-by-step development of a contemporary international migration regime.

> *Markus Kotzur is Professor of International and European Law, the Deputy Director of the Institute of International Affairs, and the Dean of Studies of the Faculty of Law at the Universität Hamburg. He is also the Director of Studies of the Master Programme "European and European Legal Studies" as well as the President of Europa-Kolleg Hamburg. Moreover, Prof. Kotzur is the Managing Director of the Institute for European Integration in Hamburg. He is also a co-editor of "Archiv des Völkerrechts". He has been as a Guest Lecturer, among others in Granada (Spain), Kolkata (Indien), Bayreuth (Germany). He is currently working on "Migration as a Challenge for International Law from a Human Rights Perspective".*

**Martin Ruhs**

### What asylum and refugee policies do Europeans want? Evidence from a cross-national conjoint experiment

*(joint paper with Anne-Marie Jeannet, Esther Ademmer, and Tobias Stöhr)*

Following the large increase in the number of migrants arriving and applying for asylum in Europe in 2015, EU member states have been engaged in highly divisive debates about how to reform the EU's asylum and refugee policies. A number of new and diverse policy models have been proposed, offering contrasting ideas about the meaning and scope of the right to asylum in Europe, resettlement of refugees from conflict regions, minimum standards of protection, assistance and cooperation with origin and transit countries, and responsibility sharing across EU member states. To debate and decide on competing visions for policy reform, it is crucial that we understand better the public's preferences for the various different dimensions of asylum and refugee policies. To this end, this paper conducts a conjoint survey experiment to study the asylum and refugee policy preferences of citizens, and how they vary across individuals, in six different EU Member States: Germany, Italy, Sweden, France, Hungary, and Poland.

> *Martin Ruhs is Chair in Migration Studies and Deputy Director of the Migration Policy Centre (MPC) at the European University Institute (EUI) in Florence. He is on leave from the University of Oxford. Martin's research focuses on the economics and politics of international migration, with a strong international comparative dimension. His books include The Price of Rights. Regulating International Labour Migration (Princeton University Press 2013) and Who Needs Migrant Workers? Labour Shortages, Immigration and Public Policy (Oxford University Press 2010, co-edited with B. Anderson). During 2009-12, he was the Director of Oxford University's Migration Observatory. During 2007-14, Martin was a member of the UK's Migration Advisory*

Committee (MAC), an independent body of five academic economists tasked to advise the UK government on labour immigration policy.

## Jeffrey H. Cohen

### A house of mirrors: refugees and reflection

Debates surrounding the refugee crisis pit xenophobes against open-mindedness. Xenophobes emphasize fear; open-minded stress tolerance. I argue that the disagreements create a house of mirrors, reflecting the position of whomever is speaking. We must embrace the facts that define refugees in place of the fantasies and look beyond the latest crisis to emphasize a shared history of mobility.

> *Jeffrey H. Cohen*, PhD is professor of anthropology at The Ohio State University. He has conducted research on migration, rural development and food in Mexico, Turkey and China with support of the National Science Foundation, National Geographic Society and Fulbright program. His recent books include, Eating Soup without a Spoon: Anthropological Theory and Method in the Real World (2015) and The Cultures of Migration: The Global Nature of Contemporary Movement (2011) written with Ibrahim Sirkeci.

## Karsten Paerregaard

### Grasping the Fear: How Migration Speaks to Anti-Globalization Sentiments and Intersects with other Controversial Issues of the Anthropocene

Why are migration, climate and AI among the most controversial issues in the contemporary world? And how do they intersect and speak to the sentiments of anger and fear that currently prevail in Western democratic countries? My talk takes its point of departure in these questions discussing how the perception of global migration has changed in the past decades and why it is now lumped together with other perils such as the natural hazards and semi-human robots. My claim is that the images of the migrant as the human Other, the climate as the natural Other, and the robot as the post-digital Other all tap into conflicting feelings of deception and blame that fuel populist movements. My conclusion is that in a time of ontological fear, migration has become an omen of where we all may end up – in a "no (hu)man's land" where only the fearless dares to walk.

> *Karsten Paerregaard* is professor of Anthropology at School of Global Studies, University of Gothenburg. His research is focused on the intersection between migration, environment, climate and culture in the Peruvian Andes. His books include Linking Separate Worlds. Urban Migrants and Rural Lives in Peru, Berg (1997); Peruvians Dispersed. A Global Ethnography of Migration, Lexington (2008); and Return to Sender. The Moral Economy of Peru's Migrant Remittances, UC Press (2015). His current book project is titled Andean Meltdown. Climate, Culture and Change in Peru.

## Ferruccio Pastore

### The "mixed migration" dilemma and the roots of Europe's migration and asylum governance crisis

Bari, March 1991: With the arrival of around 20,000 persons fleeing the chaotic Albanian transition, Europe sees, for one of the first times, a new face of international migration. Not any more guest-workers and their families, nor individual dissidents escaping persecution, but masses of disoriented fugitives not easily fitting into the mould of established legal categorisations. What will later be labelled "mixed migration" makes its appearance at the common external border of a group of countries that only a year before – with the signature of the "twin conventions" of Schengen and Dublin - had started building an international regime in the field of migration and asylum. The haphazard reactions of Italian authorities to subsequent waves of the Albanian exodus were an early warning about the genetic dysfunctions of the emerging European governance. But the warning went unobserved and only a quarter of a century later these original regulatory aporias will be fully acknowledged.

The presentation will sketch this crucial and complex history, focusing on large-scale mixed flow as a key challenge for Europe's present and future.

*Ferruccio Pastore (PhD, European University Institute, 1996) is the Director of FIERI since May 2009. He has previously been Deputy Director of the international relations and European studies think-tank CeSPI (Centre for International Policy Studies, Rome) and a post-doctoral fellow at the University of Firenze. Besides research, he has worked as an adviser on migration policy issues for Italian institutions and international organisations. He has published extensively on migration and integration policies and politics (full list on www.fieri.it).*

## Philip L. Martin

# Migration Governance in the US under Trump Administration

The US is the country of immigration, with almost 20 per cent of the world's 260 million international migrants. The number two country with international migrants, Germany, has 12 million, a fourth as many as the almost 48 million foreign-born US residents (UN DESA, 2017). The US stands alone among industrial countries in having a quarter of its immigrants, almost 11 million, unauthorised (Passel and Cohn, 2018). President Trump made reducing illegal immigration a priority. Major migration issues today include the fate of programs such as DACA, what to do about Central American families who apply for asylum, and whether to build a wall on the Mexico-US border.

*Philip L. Martin is Emeritus Professor of Agricultural Economics at the University of California, Davis and a member of the Commission on Agricultural Workers established by the Immigration Reform and Control Act of 1986. He is the author of numerous studies and reports on immigration, including Trade and Migration: NAFTA and Agriculture (1993). Prof Martin has guest-edited two issues for Migration Letters on Competitiveness in US and Japan, Migration and Development; comparing US and Mexico, and on Migration Expert Commissions.*

## Carlos Vargas Silva

# The Economic and Social Implications of Reason for Immigration

Reason for immigration (e.g. asylum, family, study, work) relates to multiple factors, such as pre-departure planning, legal rights, return plans, location choices and access to networks, which affect the economic choices of migrants and the restrictions that they face in making those choices. There is a growing body of evidence on the link between reason for migration and labour market integration, entrepreneurship, health outcomes, and remitting behaviour. This presentation discusses the latest research on these links and avenues for future research on this topic.

*Carlos Vargas Silva, is Research Director and Associate Professor at the Centre on Migration, Policy and Society (COMPAS), University of Oxford. He is also the Director of the DPhil in Migration Studies at the University. He is currently Principal Investigator for the Horizon 2020 REMINDER project about the implications of migration in Europe. Carlos is also the Principal Investigator for the ECONREF project, which explores the integration of refugees in the UK. Previously, he was Principal Investigator for the Labour Market Impacts of Forced Migration project, which explored the economic consequences of refugee migration in the Great Lakes region of Africa.*

---

In previous years, The Migration Conferences entertained many distinguished names delivering key note speeches including: Joaquin **Arango**, Complutense University of Madrid, Spain [2018], Samim **Akgönül**, Strasbourg University, France [2016], Caroline **Brettell**, Southern Methodist University, USA [2015], Pedro **Calado**, The High Commissioner for Migration, Lisbon, Portugal [2018], Barry **Chiswick**, George Washington University, USA [2014], Jeffrey H. **Cohen**, Ohio State University, USA [2012, 2015], Neli **Esipova**, Gallup World Poll, USA [2017], Nedim **Gürsel**, CNRS, CETOBaC & INALCO, France, Turkey [2015], Michelle **Leighton**, International Labour Organization, Genève,

Switzerland [2018], Philip L. **Martin**, University of California, Davis, USA [2012, 2014, 2015], Douglas S. **Massey**, Princeton University, USA [2015], Yüksel **Pazarkaya**, Novelist, Turkey [2017], Karen **Phalet**, KU Leuven, Belgium [2016], Ruba **Salih**, SOAS, University of London, UK [2018], Sasskia **Sassen**, Columbia University, USA [2017], Professor Giuseppe **Sciortino**, University of Trento, Italy [2017], Ibrahim **Sirkeci**, Regent's University London, UK [2014, 2016], Oded **Stark**, U of Bonn, Germany [2017].

## Summary Programme

### Day One – 18 June 2019, Tuesday | DAY 1
08:00 – 17:00: Registration
**08:30-09:45** – Parallel Sessions I
09:45-10:00 – Break
**10:00-12:00 – Opening Plenary Session I** [ Aula Aldo Moro, Ground Floor]
**Chairs' Welcome by** Michela C. **Pellicani**, Conference Chair, University of Bari "Aldo Moro", Italy
Opening Speech by Federico **Soda**, Director of IOM for the Mediterranean, Italy
**Moderator**: Prof Ibrahim **Sirkeci**, Regent's University London, UK
– Fiona B. **Adamson**, SOAS, University of London, UK
– Markus **Kotzur**, Universität Hamburg, Germany
– Martin **Ruhs**, European University Institute, Florence, Italy

**12:15-13:00 – Welcoming Speeches** and **Cocktail Lunch** [Atrio Centrale, Palazzo Ateneo, Ground Floor]
– Antonio Felice **Uricchio**, Rector, University of Bari "Aldo Moro", Italy
– Giuseppe **Moro**, Head of Dept. of Political Science, University of Bari "Aldo Moro"
– Michele **Emiliano**, President of the Apulia Region
– Antonio **Decaro**, Mayor of the City of Bari
– Marilisa **Magno**, Prefetto of Bari (tbc)
**13:00 – COCKTAIL LUNCH** is offered by the Municipality of Bari
[Atrio Centrale, Palazzo Ateneo, Ground Floor]
**13:00 – Launch of the Exhibition by Giovanni Perillo**
[Salone degli Affreschi, Palazzo Ateneo, 1st Floor]
**13:45-15:15** – Parallel Sessions II
15:15-15:30 – Break
**15:30-17:00** – Parallel Sessions III
17:00-17:15 – Break
**17:15-19:00 – Roundtable I: Migration Governance** [Aula Aldo Moro, Ground Floor]
– Stefano **Calabrò**, Mayor of Municipality of Sant'Alessio in Aspromonte, Italy
– Philip L. **Martin**, University of California, Davis, USA
– Martin **Ruhs**, European University Institute, Florence, Italy
**17:15-19:00** – Parallel Sessions IV
**19:15-20:00 – Concert** by the Orchestra Sinfonica della Città Metropolitana di Bari
[Atrio Centrale, Palazzo Ateneo, Ground Floor]

### 19 June 2019, Wednesday | DAY 2
08:00 – 17:00: Registration
**08:45-10:30 – Parallel Sessions V**
**08:45-10:30 – Roundtable II: I valori costituzionali alla prova del fenomeno migratorio** [Aula III - G. Di Vittorio – Ground Floor]
Chair: Raffaele **Rodio**, University of Bari, Italy
**Speakers**:
– Isabella **Loiodice**, University of Bari, Italy
– Michele **Cascione**, Lawyer, Italy
– Michela C. **Pellicani**, University of Bari, & Isabella **Piracci**, Avvocatura Generale dello Stato, Italy
– Giuseppe **Zuccaro**, Tribunal, Italy
– Vincenzo **Tondi della Mura**, University of Salento, Italy

10:30-10:45 – Break

**10:45-12:15 – Plenary Session II** [ Aula Aldo Moro, Ground Floor ]
**Moderator**: Jeffrey H. **Cohen**, Department of Anthropology, Ohio State University, USA
– Karsten **Paerregaard**, Gothenburg University, Sweden
– Carlos Vargas **Silva**, University of Oxford, UK
12:15-13:30 – Lunch
**13:30-15:00** – Parallel Sessions VI
15:00-15:15 – Break
**15:15-16:45** – Parallel Sessions VII
16:45-17:00 – Break
**17:00-18:30** – Parallel Sessions VIII
**17:30-19:15 – Roundtable III: Global Trends in Migration Policy** [Presidenza Regione Puglia, Lungomare Nazario Sauro, 33, 70121 Bari]
Chair: Michela C. **Pellicani**, University of Bari Aldo Moro, Italy
– Patrick **Doelle**, European Commission, DG Home, Italy
– Andrea Rachele **Fiore**, Puglia Regional Administration, Italy
– Gianfranco **Gadaleta**, Puglia Regional Administration, Italy
– Philip L. **Martin**, University of California, Davis, USA

## 20 June 2019, Thursday | DAY 3
**08:45-10:30** – Parallel Sessions IX
10:30-10:45 – Break
**10:45-12:30 – Plenary Session III: Reflecting on World Refugee Day** [Aula Aldo Moro]
**Chair**: Isabella **Piracci**, Avvocatura Generale dello Stato, Rome, Italy
– Antonio **Di Muro** – UNHCR representative
– Jeffrey H. **Cohen**, Ohio State University, USA
– Ferruccio **Pastore**, FIERI, Italy
– Giuseppe **Albenzio** – Vice Avvocato Generale dello Stato

**10:45-12:30 – Plenary Session IV: Technology and People on the Move** [Aula II - G. Giugni]
**Chair**: Emre Eren Korkmaz, University of Oxford, United Kingdom
– Marie **Godin** & Giorgia **Dona**, University of Oxford, United Kingdom
– Aiden **Slavin**, ID2020, United Kingdom
– Albert Ali **Salah**, Boğaziçi University, Turkey
– Zeynep Sıla **Sönmez**, UNHCR
12:30-13:30 – Lunch
**13:30-14:30 – Graduation Ceremony for Masters in Management of Migration and Integration Process**
**13:30-15:00** – Parallel Sessions X
15:00-15:15 – Break
**15:15-16:45** – Parallel Sessions XI
**17:00-18:45 – Roundtable IV: Migration Policy in Italy and Beyond** [Aula Aldo Moro]
**Chair**: Michela C. **Pellicani**, University of Bari Aldo Moro, Italy
– Massimo **Bontempi**, Central Director of the Immigration and Border Police, Home Affairs Ministry, Italy
– Luigi Maria **Vignali**, General Director for the Italians Abroad and the Migration Policies, Foreign Affairs Ministry, Italy
– Fabrice **Leggeri** (tbc), Executive Director, FRONTEX
– Jeffrey H. **Cohen**, Ohio State University, USA
– Ibrahim **Sirkeci**, Regent's University London, UK

**20:00 – 22:30: GALA DINNER at Circolo Unione - Teatro Petruzzelli**

# Day One 18 June 2019 - Registration
**Palazzo Del Prete, Foyer**

# Day One 18 June 2019 - 08:30-09:45

### Aula II - G. Giugni – Ground Floor
### 1A. Exclusion and Inclusion [D]
Chair   *Giuseppe Moro, University of Bari Aldo Moro*
44   Pilot project: Mentoring programmes for Muslim students
    Gloria Tauber
934   Brokered safety: labour recruitment and  migration assistance in the Mekong region
    Sverre Molland
683   Giant Ocean Case Reconsidered: How the fishing supply chain has transformed into an exploitation chain?
    Yi Ning Chiu & Li Chuan Liuhuang
933   Internal Migration Process and Labour Market Behaviour in Developing Countries
    Claude Sumata

### Aula III - G. Di Vittorio – Ground Floor
### 1B. Revisiting Migration Debates [D]
Chair   *Nirmala Arunasalam, University of Plymouth, United Kingdom*
599   Intercultural Relations in Georgia and Tajikistan: Post-Conflict Model
    Victoria Galyapina, Nadezhda Lebedeva, Zarina Lepshokova
256   Waves of migration in the Aegean
    İlay Romain Ors
625   European Union Citizenship. Rethinking mobile citizen's Identity
    Fulgeanu Alexandra
121   Women in the Turkish migration in France: Notes from fieldwork and discussions on the methodology
    Ecem Hasircioglu

### Aula IV - P. Labriola, Ground Floor
### 1C. Youth Migration: Unaccompanied Children [D]
Chair   *M. Gökay Özerim, Yasar University, Turkey*
154   The nexus between youthful populations and migration: The Case of MENA Region
    M. Gökay Özerim
459   Unaccompanied Teenage Refugees in Sweden: How does an underground network help and protect them?
    Amber Horning Ruf & Sara Jordeno
632   Voluntary Legal Guardianship for Unaccompanied and Separated Children in Italy: an Exploratory Research
    Francesca Viola
521   Acculturation strategies - comparison between children, adolescents and parents
    Paulina Szydlowska & Marisol Navas Luque

# Day One 18 June 2019 - 08:30-09:45

**Aula XVI - S. Fumarulo, 2nd Floor**
**1D. Outside the Boundaries of Reception**
*Chair*    *Giuliana Sanò, University of Messina; Alsos Foundation, Italy*
166    The beats of refugeeism: dwelling trajectories of a Pakistani asylum seeker in Italy, a life story approach
Sara Bonfanti
291    Out of the system. An intersectional analysis on migrants legal status and social networks
Pamela Pasian, Giulia Storato, Angela Maria Toffanin
729    Being migrant mothers: tactics, temporalities and aspirations
Selenia Marabello
175    An escapable destiny. Transgender asylum seekers' and refugees' trajectories exiting the protection system
Maria Carolina Vesce
129    Some Notes Towards an Anthropology of Homeless Asylum Seekers and Refugees
Stefano Pontiggia
488    "Always the same pain in the neck!" Living conditions of some MSM (men having sex with other men) inside and outside the asylum reception centers
Dany Carnassale

**Aula XVII - S. Bianco, 2nd Floor**
**1E. Gender and Migration**
*Chair*    *Caner Tekin, University of Bochum, Germany*
640    Interplays between Gender and other Identities? Revisiting Anti-migrant Discourses of FPÖ and Lega Nord with the Intersectionality Perspective
Caner Tekin
400    Masculinization Strategies of Male Refugees in the Process of Forced Migration
Mehmet Can Çarpar & Filiz Goktuna Yaylaci
620    Survival strategies of Nigerian victims of human trafficking in Paris
Valeria Francisca Zamorano
515    Parcours des Libanaises en situation de migration
Suzanne Menhem

**Aula XVIII - N. Calipari, 2nd Floor**
**1F. Perceptions and Migrations [D]**
*Chair*    *Tuncay Bilecen, Regent's University London, UK*
520    Pupils' attitudes towards migrants in Italy and Russia: the role of self-group and group-group distance
Lucia Bombieri
535    Gendered politics of migration management: Italian National Identity in the structuring of "others"
Fernando Aguiar, Kate Jackson, Kader A. Abderrahim
918    Security or humanity and solidarity? Populist discourse in the Visegrad Countries regarding the challenges of the migration crisis in Europe
Janusz Balicki
317    The effects of Immigration in European Public Opinion
Gökçe Bayındır Goularas & Işıl Zeynep Turkan İpek

# Day One 18 June 2019 - 08:30-09:45

**Aula XIX - F. Bravo, 2nd Floor**
**1G. Education and Migration**
Chair *Paulette K. Schuster, Hebrew University, Israel*
757  Syrian Refugees' Access to Higher Education: Mobility of Cultural Capital
Zahide Erdogan
97  First generation migrant family students on the way and within higher
education [D]
Golaleh Makrooni
1101  Teachers for Migrant and Roma Integration in Schools: the Case of North
Macedonia
Merita Zulfiu Alili, Veli Kreci, Alexander Krauss, Nada Trunk

**Aula XX - R. Fonte, 2nd Floor**
**1H. Migration and Integration**
Chair *Süreyya Sönmez Efe, University of Lincoln, United Kingdom*
595  Digital Nomads: navigating the law in their (labour) migration process
Renata Campielo
414  The role of social media in integration processes of Chinese migrants in
Austria
Carsten M. Schaefer
294  Adaptation and Acculturation: Resettling displaced tribal communities
from the protected areas in India
Madhulika Sahoo
647  Household Composition and Its Influence in the Use of School Time
amongst Latin American Migrant Teenagers in Galicia, Spain
Paula Alonso

**Aula XXI - Don Tonino Bello, 2nd Floor**
**1J. Politics and Migration**
Chair *Deniz Eroglu Utku, Trakya University, Turkey*
226  From border walls to positive policy towards forced migrants: Europe and
Israel compared
Lilach Lev Ari & Arnon Medzini
593  Denying Citizenship
Emily Ryo
466  Municipal Citizenship, State Deportation, and the International Right to
Belong
Benjamin Perryman
712  The influence of international actors in the response to the Syrian refugee
crisis in Lebanon: between humanitarian and security paradigm
Clothilde Facon

**Aula XXII – G. Di Vagno, 2nd Floor**
**1K. Work, Employment, Markets**
Chair *Ayşegül Kayaoğlu Yılmaz, Istanbul Technical University, Turkey*
485  Labour Market Effects of Syrian Refugees in Turkey:  The Case of
Informal Textile Sector in Istanbul
Ayşegül Kayaoğlu Yılmaz
501  Experiences of highly skilled Latvian emigrants in Europe and main
factors influencing their return intentions
Inese Supule

247 Middling Transnationals: Highly Skilled Migrants and the Global Commodity Chain
Shuxi Wu

773 To Move Or Not To Move? Skilled Japanese Men's Experiences of Work and Family in Australia
Iori Hamada

658 Gendered outcomes of policies towards skilled migrants: an analysis of migratory flows to Australia, Canada and the United Kingdom from 2007 to 2017
Lais Saleh

# Day One 18 June 2019 - 08:30-09:45

**Aula XXIII - T. e V.Fiore, 2nd Floor**
**1L. Identity and Migration**
Chair *Pinar Yazgan, Sakarya University, Turkey*

416 Identity Negotiation Processes and Adaptation Patterns of Gang Piao
Tianli Qin

424 Narratives of trauma among first and second generation of Pontic Greeks and their impact on diasporic identities
Georgia Lagoumitzi

509 The Gypsy Problem in Italy: the (in)visible mobility of Roma in Cagliari
Norma Baldino

18 Residents on the move: transnational opium smugglers in the interwar Balkans
Vladan Jovanovic

446 Prejudice and aesthetic preferences among skin colours
Giovanni Perillo

**Aula V. Starace, 2nd Floor**
**1M. Redes transnacionales y sociedad civil en los estudios migratorios *[SPANISH]***
Chair *Pascual García, Universidad Tecnica Particular de Loja, Mexico*

132 Migrantes venezolanos y su integración en la economía del sur del Ecuador
Carlos Fernando Soto Perez

200 Heterotopias de los migrantes del sur de Ecuador
Pablo Cañar

950 Migraciones y frontera México-Centroamérica: de frontera porosa a frontera espinosa y primer obstáculo a EE.UU
Jorge Ignacio Angulo Barredo

212 Puebla York: Migración y consumo, desde la perspectiva
Emigdio Larios Gomez

461 ¿Del sueño americano al sueño mexicano?
Mónica Patricia Toledo-González

351 El movimiento dreamer en Arizona en la era Trump
Maria Jose Enriquez-Cabral & Ismael Garcia Castro

**09:45-10:00 BREAK**

# Day One 18 June 2019 - 10:00-12:00

**10:00-12:00**

## OPENING PLENARY
**Aula Aldo Moro, Ground Floor**
Conference Chairs' Welcome
– Michela C. **Pellicani**, University of Bari Aldo Moro, Italy
Opening Speech by Federico **Soda**, Director of IOM for the
Mediterranean, Italy
**Plenary Speeches**
Chair: Ibrahim **Sirkeci**, Regent's University London, UK
– Fiona B. **Adamson**, SOAS, University of London, UK
– Martin **Ruhs**, European University Institute, Florence, Italy
  "What asylum and refugee policies do Europeans want? Evidence
  from a cross-national conjoint experiment"
– Markus **Kotzur**, Universität Hamburg, Germany
  "We, the People on the Move – the Impact of International Human
  Rights Guarantees on Migration Governance"

**12:15-13:00**

## Welcoming Speeches
**Atrio Centrale, Palazzo Ateneo, Ground Floor**
– Antonio Felice **Uricchio**, Rector, University of Bari Aldo Moro, Italy
– Giuseppe **Moro**, Head of Department of Political Science, University of
Bari Aldo Moro
– Michele **Emiliano**, President of the Apulia Region
– Antonio **Decaro**, Major of the City of Bari
– Marilisa **Magno**, Prefetto of Bari (tbc)

**13:00 – COCKTAIL LUNCH**, sponsored by the Municipality of Bari
**[Atrio Centrale, Palazzo Ateneo, Ground Floor]**

**13:00**

## Exhibition:
## Aesthetic of Migrations by Giovanni Perillo
**SALONE DEGLI AFFRESCHI, PALAZZO ATENEO, 1st Floor**

Please note the Exhibition will remain open throughout the Conference.

# Day One 18 June 2019 - 13:45-15:15

**Aula II - G. Giugni – Ground Floor**
**2A. Politiche e Strategie di Integrazione: dinamiche, regolamentazione e legittimazione dei flussi migratori in e dall'Italia [ITALIAN]**

Chair Anna Ludovici, IGOT, University of Lisbon, Portugal

268 Questioning return migration in North Eastern Italy: return policies, mobility patterns and family heritage
Melissa Blanchard

275 The new trends of Italian migration in the European scenario
Corrado Bonifazi

374 Regulatory program and empirical aspects of the integration process of immigrants in Italy
Mattia Vitiello

523 The new Italian emigration: aspects and paradoxes.
Enrico Pugliese

612 Cultural integration in urban space: between spatial segregation and transnational belonging
Anna Ludovici

**Aula III - G. Di Vittorio – Ground Floor**
**2B. Participatory Methods in Migration Research**

Chair Concepción Maiztegui-Oñate, University of Deusto, Spain

943 Exploring ways of doing research differently
Elsa Oliveira & Jo Veary

761 Adrift in a Borderland. Developing participatory and embodied methodologies as a collective of asylum seekers, refugees, activists and academic scholars
Tiina Sotkasiira, Sanna Ryynänen, Anni Rannikko, Päivikki Rapo

734 Participatory methods in migration research: challenges and possibilities
Concepción Maiztegui-Oñate

243 Study of Internal Migration India: An Application Markov Chains
Anupama Singh

**Aula IV - P. Labriola, Ground Floor**
**2C. Youth Migration**

Chair Ana Vila Freyer, Universidad Latina de México, Mexico

102 Sites of Agency and Sites of Vulnerability: Young Migrant Children"s Narrations of Inclusion or Exclusion Experiences on the Football Pitch and in the Streets of Rabat, Morocco
Chiara Massaroni

145 Who is a dreamer? The narratives of young migrants and their right to dream their lives in Mexico and the United States
Ana Vila Freyer

423 Dreams of transnational social protection: the case of Mexican policies towards youth migration
Victoria Tse

476 Migrant Youth's Multiple Logics of Migration: The Case of Japanese-Filipino Youth
Jocelyn Omandam Celero

# Day One 18 June 2019 - 13:45-15:15

### Aula XVI - S. Fumarulo, 2nd Floor
### 2D. Outside the Boundaries of Reception

Chair   *Francesco Della Puppa, Ca' Foscari University in Venice; International Center for Humanities and Social Change, Italy*

432   "We don't have a place" A case study research on mobile migrants in and outside the reception system in Northern Italy
Sebastian Benedikt

217   Homemaking and Migrant Households in Squatted Buildings: Intersecting Urban and Migration Studies
Caio Teixeira

573   Making time Everyday resistance of refugees kept on the move across the Europe
Elena Fontanari

164   Alongside the borders of reception Migrants and territories between exercise of freedom and control device
Omid Firouzi Tabar

### Aula XVII - S. Bianco, 2nd Floor
### 2E. Gender and Migration

Chair   *Joan Phillips, University of the West Indies, Barbados*

909   Sex trafficking or Migrant working

Joan Phillips

289   Precarious Migratory Status and Violence Against Women: Toward a Human Rights-Based Approach to Social Intervention with Survivors
Florence Godmaire-Duhaime

561   Risk factors for intimate partner violence (IPV) in the asylum seeking and refugee populations: a systematic review
Riyad El-Moslemany

502   Migration, gender and absolute illiteracy: looking for intersections in European social policy
Margarida Martins Barroso

### Aula XVIII - N. Calipari, 2nd Floor
### 2F. Migration and Integration

Chair   *Daniele Petrosino, University of Bari, Italy*

189   Home Food Making, Belongings and Ethnicities in the Belgian Taiwanese Migrant Women's Everyday Food Practices
Hsien-Ming Lin

675   Peer and Strength Based Approach to Refugee Integration in Seville, Spain
Virginia Paloma & Jana Sladkova

731   Migrant Trajectories and Patterns of Socialization: the case of Bulgarians in Spain
Monica Ibáñez-Angulo

343   Immigrants in the contemporary city. Urban transformation of the historical Rome's outskirt
Luca Brignone

# Day One 18 June 2019 - 13:45-15:15

**Aula XIX - L. Ferrari Bravo, 2nd Floor**
**2G. Economics of Migration**

Chair  Olgu Karan, Baskent University, Turkey

578  Analyzing Labor Market Integration Policies in EU Member States Using PROMETHEE
Anastasia Blouchoutzi, Panagiota Digkoglou, Dimitra Manou, Christos Nikas

221  Bourdieusian Multi-layered Approach in Explaining Migrant Self-Employment
Gözde İnal Cavlan & Olgu Karan

81  Integration through agricultural jobs: The Foundations in Agriculture for Refugees and Migrants (FARM) program and its impact on Canadian society and industry
Cesar Suva & Vicky Roy

369  Displaced Puerto Rican Students and Families' Experiences and Needs in Southwest Florida after Hurricane Maria
Tunde Szecsi, Debra Giambo, Hasan Aydin

**Aula XX - R. Fonte, 2nd Floor**
**2H. Migration and Acculturation**

Chair  Carlo Alberto Anzuini, University of Bari, Italy

518  The use of the relative acculturation extended model (raem) in qualitative research
Paulina Szydlowska, Jan Bazyli Klakla, Marisol Navas Luque

594  Migration, identity, and security
Liudmila Konstants

623  Social identities as predictors of intercultural attitudes
Kristina Velkova Velkova

630  Adaptation of refugees from syria and ukraine in russia in the period from 2014 to 2018 [d]
Alfiya Lyapina

361  Generalized trust as a factor of positive attitude towards migrants in Russia
Dmitrii Dubrov

**Aula XXI - Don Tonino Bello, 2nd Floor**
**2J. Identity and Migration**

Chair  H. Yaprak Civelek, Istanbul Arel University, Turkey

402  A multi-religious response to the migrant crisis in Europe: A preliminary examination of migrants' views on and experiences of multireligious cooperation on assisting their integration
Majbritt Lyck-Bowen

602  How is the integration of the Muslim community in the German context contributing to the Islamic Feminist movement?
Susana Sanchez-Muniz

714  Somali women's navigation of "Third Space" identities in online social media spaces
Samira Warsame

5  The biosocial experiences of structural inequities among immigrant and refugee communities in Atlanta, Georgia
Gregory Gullette & Marni Brown

# Day One 18 June 2019 - 13:45-15:15

### Aula XXII - G. Di Vagno, 2nd Floor
### 2K. Contextos y enfoques en los estudios migratorios *[SPANISH]*

Chair *Pascual García, Universidad Tecnica Particular de Loja, Mexico*

131 Migrant Girls on the Move: Violence and resistances in the Guatemala-Mexico-US corridor
Rodrigo Barraza

421 Del imaginario a las estrategias: Representaciones y experiencias de la ruta migratoria del Pacifico en México por los migrantes centroamericanos
Johanna Laetizia Exenberger

503 Exodus and migration in the XXIst century
Vera Magiano Hazan

61 Mexico, country for immigrant business: small-scale entrepreneurs and self-employed workers from Southern Europe
Cristobal Mendoza

239 Towards a regional migration governance in Ibero-America. Developments before the Venezuelan Diaspora (2015-2018)
Jonathan Palatz


### Aula XXIII - T. e V.Fiore, 2nd Floor
### 2L. Migration Governance in Europe

Chair *Ülkü Sezgi Sözen, Universität Hamburg, Germany*

253 The European Typology of Migration
Ülkü Sezgi Sözen

436 The political parties' approaches to the issue of immigration in Poland after 2015. Consequences for immigration policy
Monika Trojanowska-Strzaboszewska

680 More people wanted? Migration policy in shrinking regions: case study of voivodeships in Poland [D]
Kamil Matuszczyk

58 The Central Mediterranean sea between inclusion and exclusion: assembling migration management [D]
Martino Reviglio

609 The Role of The State in Migration Control [D]
Ioana Cristina Voroniuc


### Aula V. Starace, 2nd Floor
### 2M. Identity Politics

Chair *Deniz Eroğlu Utku, Trakya University, Turkey*

477 Discovering the Greek Golden Visa: An opportunity framework for migration and middlemen
Gül Üret

349 Refugee as a permanent identity: Kurdish Iranian refugees in Iraq and the never-ending road to citizenship
Dilshad Hamad Khdhir

566 Burden of Homogenization: Falling Through the Chasm of Legal Citizenship
Srija Brahmachary

220 The Impact of Muslim Migrations on Turkish Nationalism
Aslı Emine Comu

# Day One 18 June 2019 - 13:45-15:15

**Aula Aldo Moro, Ground Floor**
**2N. Il quadro dell'immigrazione nel nostro Paese: dinamiche e profili migratori (In collaboration with ISTAT - Italian Institute of Statistics)**

Chair   Ernesto Toma, University of Bari, Italy

1110   The women's economy: the job between integration and segregation
Angela Maria Digrandi

1111   Immigrants in Italy: migratory dynamics and patterns
Cinzia Conti and Fabio Massimo Rottino

1112   Quale integrazione dei lavoratori stranieri nel mercato del lavoro italiano? Un'analisi temporale tra 2008 e 2018
Michela C. Pellicani, Antonella Rotondo, Monica Carbonara, Roberto A. Palumbo

1113   L'economia delle donne e il contributo delle migranti: alcune evidenze nelle Marche e nella provincia di Macerata
Roberta Palmieri

## 15:15-15:30 BREAK

# Day One 18 June 2019 - 15:30-17:00

**Aula II - G. Giugni – Ground Floor**
**3A. Migration and Politics**

Chair   Egle Gusciute, Trinity College Dublin, Ireland

240   Italian migrants' approaches to mediated information for life-planning in Brexit times
Giuliana Tiripelli

231   Few Things I Have Learned from Them: The Representation of Genuine British People among Baltic Migrants in the UK
Martins Kaprans

431   Brexit, Immigration and Sovereignty: The Conservatives' Discourse
Sevgi Cilingir

590   The impacts of EU external migration policies on the democratization of countries in the Southern Mediterranean Neighbourhood: a case of challenge or opportunity?
Luisa Faustini Torres

**Aula III - G. Di Vittorio – Ground Floor**
**3B. Data and Methods in Migration Research**

Chair   Martina Cvajner, University of Trento, Italy

205   Boldly Go Where No Woman has Gone Before  Women Migrant Pioneers and the Quest for a New Self
Martina Cvajner

728   Fundamental Challenges in Migration Data-Access: The Case of Syrians' Fertility in Esenyurt
H. Yaprak Civelek

945   Where the Sidewalk Ends: Critical Approaches to doing Migration Research
Nergis Canefe

470   Is Documentary Research Method Sufficient to Answer Research Questions? Analyzing a Rural Locality's Remittance Investment Climate
Jeremaiah Manuel Opiniano

# Day One 18 June 2019 - 15:30-17:00

### Aula IV - P. Labriola, Ground Floor
### 3C. Youth Migration: Unaccompanied Children

Chair    *Marco Sanfilippo, University of Bari, Italy*

14    Social rights of migrant minors between the division of powers and the principle of equality. The access to nursery schools
Maria Grazia Nacci

15    The Protection of Migrant Minors under the European Convention of Human Rights
Egeria Nalin

147    Migrant minors in Europe: Between EU rules and case law of CJEU
Angela Maria Romito

362    Lost Children in the Mediterranean: Unaccompanied Refugee Minors in Turkey and Greece
Ahsen Utku

### Aula XVI - S. Fumarulo, 2nd Floor
### 3D. Outside the Boundaries of Reception

Chair    *Giuliana Sanò, University of Messina; Alsos Foundation, Italy*

152    Navigating Italy: Social networks amidst avoidance and trust for Eritrean asylum seekers
Kathryn Fredricks

339    The capability of "reception models" to resist change. The Bolognese territory after law 132/2018
Stefania Spada

143    "Vertical mobility". Migrants' trajectories within the Western Alps and beyond
Gaia Cottino

141    The floating karstic flow of asylum seekers on the North-East Italian border
Roberta Altin

### Aula XVII - S. Bianco, 2nd Floor
### 3E. Migration and Wellbeing

Chair    *Monica Ibáñez-Angulo, University of Burgos, Spain*

621    Gastroenterology Clinic for Refugees: Creative Approaches and Lessons Learned
Benjamin Levy

656    Highly skilled migration and retention of talent beyond "human capital". An analysis of the wellbeing of highly qualified migrants in Eindhoven region (The Netherlands)
Camillla Spadavecchia & Jie Yu

409    Impact of Brexit on Mental Health and Wellbeing of EU Citizens
Piotr Teodorowski

185    Perceptions, attitudes and cultural understandings of mental health in Nigeria: A scoping review of published literature [D]
Temitope Labinjo

# Day One 18 June 2019 - 15:30-17:00

**Aula XVIII - N. Calipari, 2nd Floor**
**3F. Migration and Integration**

Chair    *Erhan Kurtarır, Yildiz Technical University, Turkey*

672    Integration through Football: an Irish example
Brid Ni Chonaill

151    Border Struggles and the Autonomy of Asylum in Chios Island after the EU-Turkey Statement
Georgios (Jojo) Diakoumakos, Ioulia Mermigka, Nikos Souzas, Yannis Pechtelidis

692    Political Transition on Migration Policy in Turkey and Possibilities of Inclusiveness
Erhan Kurtarır & Elif Bali Kurtarır

433    Syrians in Turkey and Future: What Do Political Parties and Presidential Candidates Say?
Filiz Göktuna Yaylacı & Ali Faruk Yaylacı

**Aula XIX - L. Ferrari Bravo, 2nd Floor**
**3G. Economics and Migration**

Chair    *Nicola Daniele Coniglio, University of Bari, Italy*

230    Property Acquisition by Foreigners in Turkey
H. Deniz Genç & Günce Sabah Eryılmaz

592    Different motives for migration and their role for the effects of migration on household expenditure and economic activity in emigration countries
Florian Inderst

180    Promoting the Investment of Remittances for Development Policy Tools Available to Governments, Development Agencies
Maurizio Malogioglio

325    Chinese migrants in Russia in the context of the current migration and tax policy of the Russian Federation: features of business and interaction with the Russian population
Olga Zalesskaia

**Aula XX - R. Fonte, 2nd Floor**
**3H. Technology and People on the Move**

Chair    *Ruta Nimkar, Danish Refugee Council, Denmark*

562    The discursive construction of transnational migration in Internet memes. "The new diaspora" as a semioscape
Bianca Cheregi

223    Research Concept: Social Media and Smuggling Networks: What is the Link?
Ruta Nimkar, Emily Savage, Abdullah Mohammadi

357    Unaccompanied and Separated Children (UASC) and Digital Media: An exploratory research in Southern Italy
Maria Rosaria Centrone & Francesca Viola

370    Technology Assisted Overseas Migration and it's Impact of Remittance Economy in Bihar, India
Rani Kumari

# Day One 18 June 2019 - 15:30-17:00

**Aula XXI - Don Tonino Bello, 2nd Floor**
**3J. High Skilled Migration**

Chair    *Joan Phillips, University of the West Indies, Barbados*

196    Motivations for migration: Mexican middle class professionals in the United States
Lilia M Dominguez

413    Migration of Highly Skilled Specialists from Russia: Modern Features
Vladimir Iontsev & Alexander Subbotin

565    Nurses on the Move - A Gravity Model of Migration Flows
Alina Botezat

744    Contributing factors to Iranian-American students' success in the US.
Ali Reza Rezaei

**Aula XXII -G. Di Vagno, 2nd Floor**
**3K. Transnational Ties and Migration**

Chair    *Shirley Andrea Velasquez-Hoque, Oxford Brookes University, UK*

394    Ecuadorian Male Remitters in England and their Transnational Remittance Management Relations with their families in Ecuador
Shirley Andrea Velasquez-Hoque

336    Transnational activities and integration of Ukrainian migrants in Portugal and the United Kingdom
Lucinda Fonseca

491    The Senegalese community in Italy: migratory projects, transnational networks and territorial attachment
Diana Ciliberti

687    Making A New Social Space: The Transnational Activities of Syriac/Assyrians and Ezidis [D]
Ayse Guc

380    Shifted transnationalism: ethnic migrants between sending and receiving homelands [D]
Saltanat Akhmetova

**Aula XXIII - T. e V.Fiore, 2nd Floor**
**3L. Migration Law and Policy**

Chair    *Giovanni Cellamare, University of Bari, Italy*

550    The concept of unreasonable burden on the social security system of the host Member State
Solange Isabelle Maslowski

606    Scrutinizing individuals, managing flows, and the treatment of migrants
Mike Slaven

26    The evolution of family reunification in Brazilian migration policy
Luciane Benedita Duarte Pivetta

778    Between Politics and Law: strategic decision making in human displacement related humanitarian response in contemporary Brazil (2013-2019)
Joao Guilherme Casagrande Martinelli Lima Granja Xavier da Silva

# Day One 18 June 2019 - 15:30-17:00

**Aula V. Starace, 2nd Floor**
**3M. Transnational Context and Migration**
*Chair* *H. Yaprak Civelek, Arel University, Turkey*
652 Romanian emigrants and the negotiation of identities in transnational contexts
Georgiana Udrea
505 Trajectories of Steel - The making of a transnational migrant community
Dominique Santana
45 Fallouts of an Unremitting Migrant Crisis: Has Germany Lost its Identity?
Christine Anton
315 Narratives of Identity: Case Study of Middle Eastern and North African high-skilled women migrants in Japan
Isil Bayraktar

**Aula Aldo Moro, Ground Floor**
**3N. Immigration and Integration in Apulia Region**
*Chair* *Michela C. Pellicani, University of Bari, Italy*
1114 Socio-demographic characteristics of immigrant population in Apulia Region: an overall portrait
Michela C. Pellicani, Rosa Venisti, Alda Kushi
1115 The Narration of Migration in Apulia Rai News
Daniele Petrosino
1116 Perceptions over immigration: the role of geographical and social distances
Nicola D. Coniglio, Rezart Hoxhaj, Hubert Jayet
1117 Lived reception: between national model and territories. Exploratory survey on Apulian reception system
Giuseppe Campesi and Elena Carletti
1118 Best child's interests for unaccompanied migrant in regional perspectives
Annita Larissa Sciacovelli

## 17:00-17:15 BREAK

# Day One 18 June 2019 - 17:15-19:00

**Aula II - G. Giugni – Ground Floor**
**4A. Migration and Integration**
*Chair* *Monica Ibáñez-Angulo, University of Burgos, Spain*
483 The Interculturale Mediator: Resources and Competences
Simona Paula Dobrescu
299 Social capital, Acculturation attitudes, and Sociocultural adaptation of Migrants from Central Asia and South Korea in Russia
Alexander Tatarko
274 The Role of Intercultural Competence in the Relationship of Home Country Intercultural Experience and Creativity among Russian Students
Maria Bultseva
422 Multiple Identities, Acculturation, and Adjustment of Migrants from the North Caucasus in Moscow. A Network Analysis
Zarina Lepshokova & Dmitry Grigoryev

# Day One 18 June 2019 - 17:15-19:00

**Aula III - G. Di Vittorio – Ground Floor**
**4B. Göç ve Edebiyat [TURKISH]**

Chair   *Tanju İnal, Bilkent University, Turkey*
1016    Latife Tekin'in Manves City Romanında Göç Olgusu
        Tanju İnal
1018    Edebiyat Göçebesi Cemal Süreya'nın Poetikası
        Tuğrul İnal
1019    Japonya'dan Brazilya'ya 'Sōbō' Romanında Vadedilen Topraklara Göç
        Cahit Kahraman

**Aula IV - P. Labriola, Ground Floor**
**4C. Migration Law and Policy**

Chair   *Paulette K. Schuster, Hebrew University, Israel*
150     Border Externalisation and Local Tensions: The Rise of Xenophobia in
        Chios Island after the EU - Turkey Statement
        Georgios (Jojo) Diakoumakos, Nikos Souzas, Yannis Pechtelidis, Ioulia
        Mermigka
337     Responses v Responsibilities? Rethinking Localisation Trends in
        Refugee/IDP Communities
        Julie M Norman & Drew Mikhael
122     International NGOs and their political role: At the service of migration
        policies or at the service of human development?
        Paulette K. Schuster & Özge Çopuroğlu

**Aula XVI - S. Fumarulo, 2nd Floor**
**4D. Migration and Politics**

Chair   *Ayşegül Kayaoğlu Yılmaz, Istanbul Technical University, Turkey*
644     The perpetual minority: A case study of Volga German migrants in
        Germany [D]
        Anna Kozlova
23      The Scales of Attachment: Exploring territory attachment at Nahr el-Bared
        Refugee Camp and the mechanisms to overcome transience [D]
        Fabiano Sartori de Campos
474     Palestinian Refugee Policy in Postwar Lebanon: Informal Policy-Making
        through Lebanese-Palestinian Dialogue Committee (LPDC)
        Okabe Yuki
701     Contesting Crimmigration: The Migrant Day Laborer Movement and the
        Struggle for Justice
        Vanessa Guzman

# Day One 18 June 2019 - 17:15-19:00

## Aula XVII - S. Bianco, 2nd Floor
### 4E. Migration and Integration

Chair    *Ayşegül Buket Özbakır, Yildiz Technical University, Turkey*

681    Sharing the Neighborhood with Urban Refugees and Experiencing Living-Lab
Erhan Kurtarır & Ayşegül Buket Özbakır

49    Does Community Cohesion Discourse Actually Create Social Inclusion of Migrants: A Cosmopolitan Moral Analysis of the Community Cohesion Programmes in the UK
Süreyya Sönmez Efe

487    Integrating into German Society through Politics and History: The Effects of the Orientation Course (Orientierungskurs) to Refugees in Brandenburg, Germany
Wilfred Dominic Mabini Josue

516    Integration policies and training in "inclusive" societies
Ameera Masoud

713    Psychosocial and legal counseling for migrants: an analysis of services provided by university extension program in Brazil
Lisarb Valeria Montes D'oco

## Aula XVIII - N. Calipari, 2nd Floor
### 4F. Migration and Wellbeing

Chair    *Silvia Juarez-Marazzo, Chances for Children-NY, USA*

475    Health policy, migrants, refugees and asylum seekers in New Zealand
Grace Wong

302    Non-nationals' Access to Health Services in Turkey: An Assessment of Preliminary Findings
H. Deniz Genc

935    Deporting or treating? Migrant categories, health and humanitarian spaces within the context of safe migration service delivery
Sverre Molland

56    Understanding the Impact of Social Policy on Immigrants" Access to Health Care: A Theoretical Framework
Fese Elonge

## Day One 18 June 2019 - 17:15-19:15

### Roundtable I: Migration Governance

In collaboration with and support from the Municipality of Bari
**Aula Aldo Moro, Ground Floor**

Chair: Michela C. **Pellicani**, University of Bari Aldo Moro, Italy
**Speakers:**
– Stefano **Calabrò**, Mayor of the Municipality of Sant'Alessio in Aspromonte, Italy
  "L'accoglienza dei rifugiati nei piccoli comuni: un rischio o un'opportunità"
– Philip L. **Martin**, University of California, Davis, USA
  "Migration Governance in the US under Trump Administration"
– Martin **Ruhs**, European University Institute, Italy
  "Migration Advisory Committees: The UK experience"
– Matteo **Biffoni**, ANCI Delegate to Immigration Policies, Italy
  "Il ruolo dei Comuni nel modello italiano di accoglienza e integrazione: Il caso dello Sprar/Siproimi"
– Leonardo **Domenici**, President of Cittalia Fondazione ANCI, Italy
  "La specificità italiana nel contesto delle politiche europee dell'immigrazione"

17:30 - 19:15

**Concert:**
**The Orchestra Sinfonica della Città Metropolitana di Bari**
Sponsored by the Municipality of Bari
**Atrio Centrale, Palazzo Ateneo, Ground Floor**

19:15

END OF DAY ONE

# Day Two 19 June 2019 – 08:00: Registration
**Palazzo Del Prete, Foyer**

# Day Two 19 June 2019 - 08:45-10:30

**Aula II - G. Giugni – Ground Floor**
**5A. Migration Governance and Asylum Crises – The MAGYC project**

Chair   *Caroline Zickgraf (Hugo Observatory, University of Liège)*

494   Local initiatives for integration of migrants in Greece
Dimitra Manou, Anastasia Blouchoutzi, Jason Papathanasiou

504   The Invisible Majority: internal displacement and mixed migration patterns
Elizabeth Rushing

538   Navigating between hospitality and hostility. How did EU migration hotspots develop into spaces of migration governance crisis? An analysis of asylum seekers" experiences in Lesbos Island, Greece
Elodie Hut

546   European Union, Governance of Environmental Mobility and transformation of Human Rights
Camille Menu

570   Embracing and Expelling Populations: State-Building, Sovereignty and Migration Management in Greece and Turkey - A Historical Perspective on the "Migration Crisis"
Fiona B. Adamson & Gerasimos Tsourapas

**Aula III - G. Di Vittorio – Ground Floor**
**5B. Migration and Integration**

Chair   *Monica Ibáñez-Angulo, University of Burgos, Spain*

482   Acculturation profiles of ethnic Russians in Georgia and Latvia and their psychological adaptation
Tatiana Ryabichenko

527   Integration Challenges of Syrian Refugees in Multicultural Districts of Istanbul
Alev Karaduman

149   On the Adaptation and Integration in Estonia of Estonians re-migrating from Russia
Aivar Jürgenson

999   The social integration of Romanian immigrants living in Western Europe
Marius Matichescu, Daniel Luches, Alexandru Dragan

## 5C. Roundtable II: I valori costituzionali alla prova del fenomeno migratorio

In collaboration and with support of the Ordine degli Avvocati di Bari
**Aula IV - P. Labriola, Ground Floor**

Chair: *Raffaele **Rodio**, University of Bari, Italy*
**Speakers:**
– Isabella **Loiodice**, University of Bari, Italy
   "Flussi migratori tra Europa dei diritti e Europa dei doveri"
– Michele **Cascione**, Lawyer, Italy
   "Verso la disciplina di una migrazione sostenibile"
– Michela C. **Pellicani**, University of Bari, & Isabella **Piracci**, Avvocatura General dello Stato, Italy
   "Diritto di cittadinanza: quali scenari possibili"
– Giuseppe **Zuccaro**, Tribunal, Italy
   "Fenomeno migratorio e lavoro sommerso"
– Vincenzo **Tondi Della Mura**, University of Salento, Italy
   "Sicurezza, dignità umana e solidarietà. La regolamentazione del fenomeno migratorio alla luce dei valori costituzionali"

*Left margin: 08:45 - 10:30*

# Day Two 19 June 2019 - 08:45-10:30

### Aula XVI - S. Fumarulo, 2nd Floor
### 5D. Prospects in Migration Theory
Chair   Carmen Caruso, Regent's University London, United Kingdom
203   Drivers of human migration - a review of scientific evidence
Dino Pitoski, Thomas Lampoltshammer, Peter Parycek
262   Migration and security: theoretical issues
Evgenii Vyasheslavovich Gamerman
181   Is migration a threat or an opportunity for both origin and destination countries? Policies to harness the potential of economic migration
Maurizio Malogioglio
428   Towards an Integration of Models of Discrimination of Immigrants: From Ultimate (Functional) to Proximate (Sociofunctional) Explanations
Dmitry Grigoryev & Anastasia Batkhina

### Aula XVII - S. Bianco, 2nd Floor
### 5E. Gender and Migration
Chair   Martina Cvajner, University of Trento, Italy
736   Revisiting Gender (Stereotypes) in Understanding the Migrant Experience
Christine Inglis
206   Uncanny Babushka Migration and  the Search for a new Sexual Self
Martina Cvajner & Giuseppe Sciortino
628   Spaces of Identity: The role of the Marxloh Mosque in Shaping Turkish-German Women's Integration and Sense of Community
Irem Oz
213   Protecting Women Victims of Human Trafficking In Mixed Migration Flows: the Challenge of Identification in Italy and Greece
Francesca Cimino & Paola Degani

### Aula XVIII - N. Calipari, 2nd Floor
### 5F. Turkey's Migration Space
Chair   Caner Tekin, University of Bochum, Germany
637   Refugee Studies on Turkey: The Current State of the Art
Caner Tekin
735   Exploring coping strategies of urban refugee women in Iskenderun, Turkey
Maria Kanal
544   Representation of Kurdish Refugees in Japanese Media
Vakkas Çolak
639   Illegal Immigration as a Struggle to Hold onto Life: The Example of Istanbul
Celal Altin

# Day Two 19 June 2019 - 08:45-10:30

**Aula XIX - F. Bravo, 2nd Floor**
**5G. Citizenship Debates**
Chair   *Deniz Eroğlu Utku, Trakya University, Turkey*
395     Implementation of EU migration rules in the Czech Republic: migration quotas and beyond
        Vaclav Stehlik
582     Discrimination and Bottom-Up Nationalism on Social Media. An Analysis of the Citizenship Debate in Turkey
        Cigdem Bozdağ
740     Defining Migrant Citizenship in Urban China: Paradigm Shift and Context Contingencies
        Xiaorong Gu
 16     Citizenship as a Commodity? A Critical Analysis of Citizenship Literature and Immigrant's Conceptualization of Citizenship [VIDEO]
        Deniz Yetkin Aker
298     International Relations and Refugee Policies: A comparative Study of Iran and Turkey
        Sussan Siavoshi

**Aula XX - R. Fonte, 2nd Floor**
**5H. Migration and Wellbeing**
Chair   *Silvia Juarez-Marazzo, Chances for Children-NY, USA*
110     Barriers to Health Care Access and Service Utilization of Refugees in Austria: Evidence from a Cross-Sectional Survey
        Isabella Buber-Ennser
138     The Adverse Health Effects of Punitive Immigration Laws in the United States
        Nicholas A. Vernice & Lisa V. Adams
674     The Italian "Diciotti" vessel case, out of political disputes: state of health of rescued Eritrean migrants
        Giancarlo Ceccarelli, Serena Vita, Maurizio Lopalco, Silvia Angeletti
596     The Role of Multiple Identities and Acculturation Strategies in Psychological Wellbeing of Crimean Tatars
        Ekaterina Kodja

**Aula XXI - Don Tonino Bello, 2nd Floor**
**5J. Migration Policies in Europe**
Chair   *Ülkü Sezgi Sözen, Universität Hamburg, Germany*
183     "Sand in the machinery" - A comparative analysis of political responses to unaccompanied afghan minors seeking asylum in Scandinavia
        Marianne Garvik
902     Public goods theory and refugee protection in the EU: political parties perceptions
        Diego Caballero Vélez
384     Lived Realities and Survival Tactics of Undocumented Migrants in Finland in the Context of Tightened Asylum Policies
        Abdirashid Ismail
654     The migration and refugee crisis in Europe: Poland case study [D]
        Marta Pachocka, Dominik Wach

# Day Two 19 June 2019 - 08:45-10:30

## Aula XXII -G. Di Vagno, 2nd Floor
### 5K. Identity and Migration
Chair   *Hande Erdem-Möbius, University of Berlin, Germany*
537   Turkish Immigrant Background Mothers' Perceptions on the Use of Mother Tongue
     Hande Erdem-Möbius, Özen Odağ, Yvonne Anders
743   Migrants' taking on perceptions of identity
     Nadja Stamselberg
232   Ethnic Habitus, Moral Judgments and Un/deserving Migrants: Intimate apartheid inside an asylum seeker day center
     Jenae Carpenter
510   Language learning, integration and identity: the experiences of adult migrant and refugee learners of English in Britain
     Jill Court
484   Perceptions and Aspirations of Syrians in Turkey
     Aysegul Kayaoglu

## Aula XXIII - T. e V.Fiore, 2nd Floor
### 5L. Retos y oportunidades en la migracion internacional *[SPANISH]*
Chair   *Pascual García, Universidad Tecnica Particular de Loja, Mexico*
124   Retropía: un anhelo del conservadurismo nativista
     Laura Natalia Rodríguez-Ariano
915   Migración y resiliencia: un estudio empírico sobre la familia transnacional de retorno
     José Salvador Cueto Calderón, Ismael García Castro & Nayeli Burgueño Angulo
916   La difícil gestión de la migración de tránsito por México. Efectos locales de problemáticas globales
     Mirko Marzadro & Ismael García Castro
295   El retorno desde España a Uruguay
     Martin Koolhaas
170   El muro fronterizo entre México y Estados Unidos. Imaginarios sociales sobre los límites territoriales y la migración laboral irregular (1994-2018)
     Migue Aguilar Dorado & Ana Laura González
419   Expelled and integrated? The migration of elders
     Pascual García

## Aula V. Starace, 2nd Floor
### 5M. Göç ve Siyaset [TURKISH] [D]
Chair   *Emine Güzel, Tekirdağ Namık Kemal University, Turkey*
1012   Siyasi Yozlaşmanın Göç Üzerindeki Etkileri
     Hanife İnan, Evindar Güneş
1013   Göçün Kültürel, Toplumsal ve Mekansal Yapı Üzerinde Etkileri
     Berke Özdemir
1014   Göçmenlerin Bulundukları Ülkelerdeki Demokrasilere Siyasi Katılım Yolları
     Muhammed Tayyip Tellioğlu, Fatma Nur Magunacı
1015   Doğa Toplumundan Siyasal Topluma, John Locke'a Eleştirel Bakışın Göçle Konumlandırılması
     Hatice Menteşe, Duygu Akgül
234   Göçmenlere Yönelik Sosyal Destek Grupları
     Sevim Atila Demir & Pınar Yazgan

## Day Two 19 June 2019 - 08:45-10:30

**Aula Aldo Moro, Ground Floor**
**5N. Policies of Refuge around the World**

Chair *Elli Heikkila, Finland Institute of Migration, Finland*

720 Refugee Status Determination Policy and Practice: will the current "zero boats" status be sustained by the newly proposed policy of the Australian Labor Party?
Petra Madge Playfair

967 Japanese Refugee Policies and the Syrian Asylum Seekers in Japan
Yahya Almasri

765 The situations of Stateless and Migrants in Thailand
Darunee Paisanpanichkul

305 Criminalization of Foreign Bodies in Japan: Legitimatization of Strict Control of Foreign Workers, Asylum Seekers, and Refugees
Ayako Sahara-Kaneko

**10:30-10:45 – BREAK**

## Plenary Session II

**Aula Aldo Moro, Ground Floor**

Chair: Professor **Jeffrey H. Cohen**, Ohio State University, USA

**10:45-12:15**

**Speakers:**
– Professor **Karsten Paerregaard**, Gothenburg University, Sweden
"Grasping the Fear: How Migration Speaks to Anti-Globalization Sentiments and Intersects with other Controversial Issues of the Anthropocene"
– Dr **Carlos Vargas Silva**, University of Oxford, UK
"The Economic and Social Implications of Reason for Immigration"

**12:15-13:30 – Lunch**

## Day Two 19 June 2019 - 12:15-13:30 MOVIE SCREENING

946

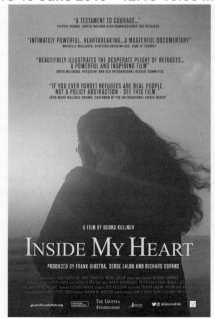

**Aula Aldo Moro, Ground Floor**

*INSIDE MY HEART*
*A FILM BY*
*DEBRA KELLNER*

# Day Two 19 June 2019 - 13:30-15:00

### Aula II - G. Giugni – Ground Floor
### 6A. Prospects in Migration Theory
Chair   *Giuseppe Sciortino, University of Trento, Italy*
385   Two Paradigms? Migration, Mobility and Social Theory
Giuseppe Sciortino & Nicholas D. Harney
216   The applicability of migration theories in the context of South-South migrations
Julieta Bengochea
618   Migration and cleavage theory
Daniele Petrosino
917   Long and short-distance migration motivations in Namibia: A gravity model approach
Eldridge Moses

### Aula III - G. Di Vittorio – Ground Floor
### 6B. Contextos y enfoques en los estudios migratorios *[SPANISH]*
Chair   *Pascual García, Universidad Tecnica Particular de Loja, Mexico*
547   Enfoques interpretativos sobre desplazamiento forzado
Yuber Hernando Rojas Ariza & Ledis Bohorquez Farfán
629   Narrativas de migrantes en contextos transnacionales
Diana Tamara Martinez Ruiz & Alejandra Ceja Fernandez
738   Haciendo Antropologia de la migracion en contextos
Monica Ortiz Cobo and Rosella Bianco
452   Precariedad laboral en ecuador: Acua¡l es el rol de las remesas y el crecimiento economico?
Carlos A. Moreno Hurtado, Ximena Songor Jaramillo
427   Gentrificación transnacional: Comunidades de migrantes
Ismael Garcia Castro, Miguel A. N. Higuera, Claudia Canobbio Rojas
300   Desde y hacia la migración "ordenada". Multilateralismo y capacity-building como motores de las políticas de control migratorio global
Silvana Estefania Santi Pereyra

### Aula IV - P. Labriola, Ground Floor
### 6C. Migration and Policy in the South
Chair   *Simeon Magliveras, King Fahd University of Petroleum and Minerals, Saudi Arabia*
251   A Comparison of Migration Regulatory Laws of Saudi Arabia And Bahrain: More Equitable Livelihoods for Immigrants to the GCC?
Simeon Magliveras
123   Making diaspora policies without knowing the diaspora: The case of Sri Lanka
Pavithra Jayawardena
66   Structuration of International Migration: Historicizing Overseas Migration from a Philippine Rural Province of Origin
Violeta Alonzo, Elizabeth Marfel Gagni, Ana Marie Faith Corpuz, Jeremaiah Manuel Opiniano
161   Defying the Sending-State in the Gulf: How Do Transnational Filipino Migrants with Disabilities Use Migration to Access Local and Global Employment Barriers?
Froilan Tuccat Malit Jr

# Day Two 19 June 2019 - 13:30-15:00

### Aula XVI - S. Fumarulo, 2nd Floor
### 6D. Migration and Identity

Chair   *Evangelia Tomara, City University, London, UK*

960   Refugees Complex Identities in Times of Hostility
Rosella Bianco & Monica Ortiz Cobo

225   Where I Belong
Sanjana Ragudaran

777   Hot Hate Waves, Summer 2018 in Germany - Is this a sign of social climate change in Germany?
Asiye Kaya

670   Am I German, am I Turkish, or both? Ethnic identity construction of third-generation people with Turkish background in Germany
Mladen Adamovic & Marina Adamovic


### Aula XVII - S. Bianco, 2nd Floor
### 6E. Identity and Migration

Chair   *Rafaela Pascoal, University of Padua, Italy*

20   The inherent contradictions of the case worker and the welcoming system for asylum seekers: an ethnographic study of the Autonomous Province of Trento, Italy
Elena Giacomelli

108   Migrants' attitudes towards migration and other migrants
Ilona Bontenbal

4   Personal value preferences and immigrant group appraisal as predictors of voluntary contacts with immigrants among university students
Eugene Tartakovsky

450   Trafficking in human beings between a gender-based and a gender-biased approach: Exploring the failure to identify adult male victims
Noemi Magugliani


### Aula XVIII - N. Calipari, 2nd Floor
### 6F. Work, Employment, Markets

Chair   *Elli Heikkila, Finland Institute of Migration, Finland*

233   Immigrants and the Finnish labour market and future prospects
Elli Heikkila

926   Workers from the Northern Triangle of Central America (NTCA) in the Mexican Labor Market
Liliana Meza & Carla Pederzini

320   European high-skilled workers: what contribution to Brussels socio-spatial divisions?
Charlotte Casier

304   Abject Zones and Parlous Mobilities: Understanding Migrant Precarity and Spaces of Non-existence in the Narratives of Ilocano Transit Migrants in Selected Gulf Countries
Ana Marie Faith Laeno Corpuz

# Day Two 19 June 2019 - 13:30-15:00

### Aula XIX - L. Ferrari Bravo, 2nd Floor
### 6G. Arts and Migration

Chair   Tunde Szecsi, Florida Gulf Coast University, USA
506   Representation of migrant women in New Turkish Cinema
      Deniz Bayrakdar
293   "A world stripped of all borders". Re-framing Identity in Thomas Arslan's
      Berlin Trilogy
      Luisa Afonso Soares
327   Integration policies and collaborative creation through the audiovisual
      sector: the Milanese project "Cinema di Ringhiera"
      Maria Francesca Piredda
250   Migration and the Motherland: Ex-Colonisers and Ex-Colonised in the
      literary works of Jose Eduardo Agualusa and Ondjaki
      Maria Backe

### Aula XX - R. Fonte, 2nd Floor
### 6H. CROAS: Migration and Professional Social Service *[ITALIAN]*
In collaboration with CROAS Puglia

Chair   Patrizia Calefato, University of Bari, Italy
976   Professional social service and immigration. The results of the first
      national research
      Patrizia Marzo
973   The Social Assistant in the UAMS system. The PUERI project
      Antonio Brascia
974   Problematic nodes and strengths of the multidisciplinary teams
      Sabrina Callea
975   Functions and competences of the S.A. working with adult immigrants
      Ignazio Galeone

### Aula XXI - Don Tonino Bello, 2nd Floor
### 6J. Türkiye'de Göç ve Uyum *[TURKISH]*

Chair   Gökçe Bayındır Goularas, Yeditepe University, Turkey
318   Sınır Duvarları
      Gökçe Bayındır Goularas & Nihan Kocaman
534   Bir Ülkeye Dışarıdan Bakmak: Sovyetler Birliği Gezilerinde Öznellik
      Tuncay Bilecen
922   Türkiye'de Göç Hareketliliği Entegrasyon ve Mübadele
      Sibel Terzioğlu

### Aula XXII -G. Di Vagno, 2nd Floor
### 6K. Push for Migration

Chair   Nicola Daniele Coniglio, University of Bari, Italy
120   Adapting on their own terms: the myth of climate migration in small island
      developing states
      Caroline Ferguson & Autumn Bordner
456   Internal Displaced People, social unrest and mobilisations: a convenient
      outcome?
      Yolanda Hernandez-Albujar & Alfonso Sanchez Carrasco
117   Escape Mobilities: Rethinking the autonomy of Migration
      Ayushman Bhagat

## Day Two 19 June 2019 - 13:30-15:00

**Aula XXIII - T. e V.Fiore, 2nd Floor**
**6L. Migration and Integration**

Chair   Ilkay Sudas, Ege University, Turkey
272   Urban Related Identity among Turkish Immigrants In Germany and The Netherlands
Ilkay Südas
330   Where is the emotional home of Turkish migrants in Germany?
Haci Halil Uslucan
237   The Lived Experience of an Integration Paradox: An In-Depth Exploration of Structurally Successful Migrants' Struggle with Belonging
Nella Geurts, Tine Davids, Niels Spierings
215   Social Capital and Migrant Integration in Greece
Dimitrios Georgiadis & Maria Korizi

**Aula V. Starace, 2nd Floor**
**6M. Migration and Integration**

Chair   Ana Vila Freyer, Universidad Latina de México, Mexico
308   School-aged immigrant bilinguals' language use in multiple social settings: a case study of Chinese-Australian children
Yilu Yang
333   All welcome here? Discrimination in the Irish housing market
Egle Gusciute
368   Precarious Mobility, Vulnerability, and Precarity: Ethnographic evidence from the Mising Youths of Assam
Bipul Pegu
381   Educational paths of migrant origin young people in Nordic societies
Maili Malin & Andrea Dunvaly
94   Educational Achievement among U.S. Undocumented College Students
Nicole Kreisberg

## Day Two 19 June 2019 - 13:30-15:00 VIDEO SCREENING

**Aula Aldo Moro, Ground Floor**

### *Migrant landings in Apulia region: Red Cross in action*
### *A SHORT VIDEO BY RED CROSS BARI COMMITTEE*

**15:00-15:15 BREAK**

## Day Two 19 June 2019 - 15:15-16:45

**Aula II - G. Giugni – Ground Floor**
**7A. Migration and Integration**

Chair   Liudmila Konstants, American University in Central Asia, Kyrgyzstan
198   Migration and the Classroom: A Comparative Study of Education Practices for Supporting New Arrivals in Three Communities in Canada and US
Kristen Nielsen
392   "You have to do the extra work" Young Finnish-Russian dual citizens and conflicting perceptions of Russia
Marko Kananen
111   Unscrupulous Intermediaries in International Labour Migration: Nepal
Anurag Devkota
281   Economic and Social Impact of International Youth Migration and Unemployment
Solomon Kunle

# Day Two 19 June 2019 - 15:15-16:45

## Aula III - G. Di Vittorio – Ground Floor
### 7B. Prospects in Migration Theory

Chair   Carmen Caruso, Regent's University London, United Kingdom

968   Micro Level Approach to Conflict Model of Migration: Intersectional analysis of Perception
Pınar Yazgan and Deniz Eroğlu Utku

340   Operationalizing the Willingness to Migrate through the Capability Approach
Chinedu Obi, Fabio Bartolini, Marijke D'Haese

442   What is left unsaid: intersectionality, public space, and the Syrian diaspora in London
Carmen Caruso

921   Chinese rural-urban migrant integration and urban public space: A theoretical framework
Chen Qu

## Aula IV - P. Labriola, Ground Floor
### 7C. Gender and Migration

Chair   Margarida Barroso, ISCTE-IUL, Portugal

568   Comparing Syrian Female Urban Refugees' Experiences in Bursa, Turkey: A Socio-spatial Perspective at The Intersection of Class, Gender, Ethnicity and Urban Space
Gulcin Tunc

913   Marriage, mobility and migration: reflections on gender relations in Nigerian society
Paula Morgado

914   Gender based Violence among "Semi-Moroccan Migrant Families": Migration between a Liberating Phenomenon and a Restricting one in Najat El Hachmi's The Last Patriarch
Wissam Bitari

563   Women and migrants. Identities and perceptions in the reception system
Clara Vecchiato

## Aula XVI - S. Fumarulo, 2nd Floor
### 7D. Onward Migration in a Changing Europe

Chair   Francesco Della Puppa, Ca' Foscari University in Venice, Italy

128   For the sake of our children: The role of children's educational trajectories in decisions of Somali families to move onward within the EU
Marloes de Hoon & Ilse van Liempt

321   Familial strategies for getting by: negotiating citizenship and mobility in times of crisis
Helen McCarthy

931   The Social Trajectories, Material Conditions and Geographic Mobility of Asylum Seekers and Beneficiaries Outside the Italian System of International Protection
Giuliana Sanò & Francesco Della Puppa

176   Placing the future: remigration, citizenship and relatedness among London's Portuguese-Bangladeshis
Jose Mapril

# Day Two 19 June 2019 - 15:15-16:45

### Aula XVII - S. Bianco, 2nd Floor
### 7E. Edebiyat ve Göç [TURKISH]
Chair   Emine Güzel, Namik Kemal University, Turkey
1017   Zorunlu Göçe Yeni Bir Yaklaşım: Ötenazi
Emine Güzel
119   Çerkes Sürgününün Edebiyata Farklı Yansımaları

Gülnihal Gülmez
533   Mübadele Romanlarında Anadili Farklı Olan Grupların Durumu
Hülya Bayrak Akyıldız

### Aula XVIII - N. Calipari, 2nd Floor
### 7F. Migration and Integration
Chair   Karolina Nikielska-Sekula, University College of Southeast Norway
499   Home as continuum. Home making in a context of multigenerational diasporic communities
Karolina Nikielska-Sekula
667   The (im)mobilities of Japanese lifestyle migrants in Austria and Bulgaria: Agency vs. structural determinants
Yana Yovcheva
651   Social Protection on the Move: A Transnational Exploration of Nicaraguan Migrant Women's Engagement with Social Protection in Spain and Nicaragua
Chandreyi Guadalupe Guharay
586   The migrants' journey as the ground for an anticipatory socialization to the reception system
Elena Carletti

### Aula XIX - L. Ferrari Bravo, 2nd Floor
### 7G. Migration and Power
Chair   Jennifer Lopes, University of Montreal, Canada
284   The Regional Disembarkation Arrangements as a new (?) tool to externalize the migration issue: a "marriage of convenience" between EU Member States and third countries
Annalisa Geraci
904   The Impact of Current and Emerging US Immigration and Refugee Law on Human Rights, Social Justice, and Civil Liberty
John Thomas
63   The Unmitigated Harms of US Immigration Detention for Noncitizens
Jennifer A. Bulcock
627   Relations Between Migration and Power: The Case of Power Transition Theory
Fazli Dogan & Erdem Özlük

# Day Two 19 June 2019 - 15:15-16:45

**Aula XX - R. Fonte, 2nd Floor**
**7H. Work, Employment, Markets**

Chair   Liliana Meza Gonzalez, National Institute of Statistics and Geography, Mexico

202   A Comparative Study to Examining the Entrepreneurial Activity: The Case Turkish Cypriot, and Mainland Turkish and Kurdish Habitus in Small Business Ownership in the Food Catering and Retail Sectors in London
Olgu Karan

31   Assessment of Skilled Immigrants in South Africa: An Overview of Small Business Activities in Cape Town
Emmanuel Ilori

112   Romanian returnees" entrepreneurship: challenging the idealized view of the opportunity driven entrepreneurship
Alin Croitoru

636   Performance and social mobility of migrant entrepreneurs: resources and individual trajectories
Daniela Gnarini

**Aula XXI - Don Tonino Bello, 2nd Floor**
**7J. Türkiye'de Göç ve Uyum [TURKISH]**

Chair   Mustafa Yakar, Süleyman Demirel University, Turkey

579   Kentsel Mültecilerin Barınma Deneyiminin Adaptasyon Süreçlerine Etkisi: Bir Sınır Kenti Hatay Örneği
Zehra Güngördü & Erhan Kurtarır

949   Türkiye'de Yaşlılığın Mekânsal Görünümünü Yaşli Göçleri "Nerede", "Nasil" Etkiliyor?
Mustafa Yakar, Kadir Temurçin, İsmail Kervankiran

453   Türkiye'de Mültecilerin Sağlık Hakkına Erişimi
Kardelen Coşkun

634   Mardin ilindeki en korumasız kişiler için entegre koruma hizmetlerinin sağlanması: Bütüncül Koruma Projesi örneği
Leyla Onur Yanar

435   Tarihi ve Kültürel Bağların Uyum Üzerindeki Etkisi: Hatay'daki Geçici Koruma Kapsamındaki Suriyeliler
Fırat Mustafa Tekbaş

**Aula XXII -G. Di Vagno, 2nd Floor**
**7K. Migration and Politics**

Chair   Ülkü Sezgi Sözen, Universität Hamburg, Germany

700   Devising Depravity: Examining Categorical Imperatives and Approaches to Morality in Immigration Jurisprudence
Abel Rodriguez

969   An investigation into how greater harmonization of the decision making process in the Common European Asylum System could help achieve greater fairness and efficiency
Matthew Bodycombe

512   The Obligation of Businesses to Protect the Rights of People Potentially Affected by Development-induced Displacement: the UN Guiding Principles on Business and Human Rights in Context
Roman Girma Teshome

# Day Two 19 June 2019 - 15:15-16:45

**Aula XXIII - T. e V.Fiore, 2nd Floor**
**7L. Question of "Crisis"**

Chair   Inci Aksu Kargin, Usak University, Turkey
187     Transit and Asylum in Italy: Complicating the Figure of the "Economic Migrant" in Contexts of Crisis
        Eleanor Paynter
603     Humiliation in the Refugee Crisis: A study of refugee testimony
        Adrienne de Ruiter
118     Getting to the Root of the International Migrant Crisis: What Moved the Asylum Seekers in Germany to Flee their Country
        Christina Kohler, Carlos Denner dos Santos, Marcel Bursztyn
57      Moving from Crisis Discourse to Everyday Practice
        Claudia Lintner

**Aula V. Starace, 2nd Floor**
**7M. Refugee Policies and Reflections**

Chair   Esin Bozkurt, GIZ, CLIP Project, Turkey
480     Leave no one behind: The CLIP Experience in Turkey
        Esin Bozkurt
373     The Path Forward: A Humane Solution for El Salvadorians with TPS
        Zaana Malia Hall
545     The No-Law Land: The Consequences of the Italian-Libyan Agreements on Refugees
        Rosa Aloisi
77      Reformulation of Colonialism and Its Effects on Refugees: The Comparative Case of Syrian Refugees in Turkey, Greece and Germany
        Baris Oktem

**Aula Aldo Moro, Ground Floor**
**7N. Narratives of Migration**

Chair   Olivia Joseph-Aluko, Reinvent African Diaspora Network, UK
188     African migration to Europe [D]
        Olivia Joseph-Aluko
19      "We will meet at the Bridge": Alexander Bridge and stories of Zimbabweans[D]
        Thembani Dube
420     Latino communities in Sao Paulo (Brazil) and their webdiasporic communication
        Camila Escudero, Mohammed ElHajji
75      Marching for identity, recognition and visibility: Sikh processions in the Metropolitan Area of Barcelona
        Nachatter Singh Garha

**Aula V. Buono, 6th Floor**
**7P. Children on the Move**

Chair   Ana Vila Freyer, Universidad Latina de México, Mexico
638     Access to justice for migrant children detained in Canada
        Jennifer Lopes
383     Labor Migration and Deportability of "Illegal"-Resident Children: A Comparative Analysis of the State of Israel and South Korea [D]
        Yuri Keum
607     The well-being of children of CEE "liquid migrants"
        Daina Grosa
583     Children of Syrian Refugees and Their Search for a Home: A Case of Childhood Statelessness
        Vartika Anand

**16:45-17:00 BREAK**

# Day Two 19 June 2019 - 17:00-18:45

### Aula II - G. Giugni – Ground Floor
### 8A. Migration and Integration

Chair   *Olgu Karan, Baskent University, Turkey*
412     Intercultural Relations in Georgia and Tajikistan: Post-Conflict Model
        Nadezhda M Lebedeva, Victoria N Galyapina, Zarina Kh Lepshokova,
        Tatiana A. Ryabichenko
635     The relationship between demographic variables, perceived
        discrimination and perceived stress in an African international student
        population at the University of the Western Cape, South Africa
        Faranha Isaacs
411     A country built on the paper: "Italiani pel mondo"
        Gaetano Morese

### Aula III - G. Di Vittorio – Ground Floor
### 8B. Remittances

Chair   *Farid Makhlouf, University of Pau, France*
246     Do Remittances strengthen Trust in the Home Countries? A Quick Look at
        the Data
        Farid Makhlouf
292     The Role of International Remittances in Poverty Reduction in Resource-
        Poor Former Soviet Republics
        Sarvar Gurbanov, Vusal Mammadrzayev, Hasan Isgandar
 90     Impact of Remittance on socio-economic development in the village of
        Tarai/Madhesh, Nepal
        Deepak Chaudhary
472     Behavioural Economics of Migration? An Exploratory Framework on the
        Local Development Potential of Overseas Remittances
        Jeremaiah Manuel Opiniano
278     Understanding relationships and remittance flow during the migration
        period: Strength of social ties as a factor determining remittance
        behaviour
        Thebeth Rufaro Mukwembi

### Aula IV - P. Labriola, Ground Floor
### 8C. Space Reconsidered

Chair   *Ayşegül Özbakır, Yildiz Technical University, Turkey*
653     Place-Making Practices of Transnational Immigrants
        Meltem Oktay, Erhan Kurtarır, Ayşegül Özbakır
677     The digital space as means of production of intimacy in refugees under
        the European Commission's Relocation Scheme
        Marta Lemos
953     The role of place-based factors in the international migration of health
        workers – experiences from Hungary
        Szabolcs Fabula, Viktor Pál, Zsófia Ilcsikné Makra, Gábor Lados

# Day Two 19 June 2019 - 17:00-18:45

### Aula XVI - S. Fumarulo, 2nd Floor
### 8D. Gender and Migration

Chair    *Gul Ince Beqo, Catholic University of Milan, Italy*

109    Gender roles and decision-making in the migration from Turkey to Italy
Gul Ince Beqo

156    Women's liberation from gender-based violence: The perspective of migrant-origin women in South Tyrol (Northern Italy)
Marina Della Rocca, Dorothy Louise Zinn

195    The Power of The Homeland: Gender Roles and the Transnational Ripple Effect among Liberian Refugees in the Diaspora
Bernadette Ludwig

### Aula XVII - S. Bianco, 2nd Floor
### 8E. Migration Experiences and Coping Strategies

Chair    *Vildan Mahmutoglu, Galatasaray University, Turkey*

211    Youtube: The Virtual Communication Between Poblano Migrants (De Puebla York) and Their Families in Mexico (Puebla)
Emigdio Larios-Gomez

549    Male Out-Migration and the Left-Behind Women in Rural Odisha
Disha Das

169    Perception of 1864 Migration in Public Memory
Melike Batgıray

920    Zimbabwean Parents' experiences of bearing and raising children in the UK
Ruvimbo Machaka

552    Strategies to cope with the negative social identity: A social psychological study on African immigrants in Istanbul [D]
Yakup Azak, Burcu Tüm, Tuba Aydın

### Aula XVIII - N. Calipari, 2nd Floor
### 8F. Work, Employment, Markets

Chair    *Daniele Petrosino, University of Bari, Italy*

517    Does gender has a significant role in reintegration of return migrants? Evidence from gulf returnees in rural Kerala, India
Reshmi R S, Afsal K, Sayeed Unisa

158    Migration-induced stressors and coping responses among labour migrants - a qualitative study among returnee migrants in Nepal
Joelle Mak

486    The difficulties of refugees in the labour market: a network approach
Michael Eve & Maria Perino

105    Workplace Segregation and Occupational Segmentation: Evidence from Estonian, Swedish and Russian Immigrants in Helsinki Metropolitan Area
Anastasia Sinitsyna

686    Young *Au Pairs* living in France: Another face of domestic exploitation?
Miriam Alejandra Martinez Torres

## Day Two 19 June 2019 - 17:00-18:45

**Aula XIX - L. Ferrari Bravo, 2nd Floor**
**8G. Intersectionalities and critical geographies of mobility: multiscalar insights in the Iberoamerican region [D]**

Chair    *Ana Santamarina Guerrero, University of Glasgow, UK*
*Almudena Cabezas Gonzalez Universidad Complutense de Madrid, Spain*

664    Becoming a mass emigration country, An attempt to measure outflow of Venezuelans worldwide
Thais Garcia Pereiro & Victoria Prieto

490    Disrupting structural racism in Madrid city: challenges of migrant solidarity politics
Ana Santamarina Guerrero

663    Mujeres colombianas negras y pacíficas cruzando las fronteras
Maria Margarita Echeverry

376    [Not just] people on the move. An intersectional and multiscalar analysis of Colombian migration
Paula Medina García

608    Intersectionality and refugees. Good practice cases in Lisbon
Giovanna Gonzalez

398    Gender political activism contesting European racialised border in Madrid
Almudena Cabezas Gonzalez

## Day Two 19 June 2019 - 17:30-19:15

### Roundtable III: Global Trends in Migration Policy

REGIONE PUGLIA

In collaboration with and support from the Puglia Regional Administration

**Sala S. Nicola, Presidenza Regione Puglia, Lungomare Nazario Sauro, 33, 70121 Bari**

Chair: Prof Michela C. **Pellicani**, University of Bari Aldo Moro, Italy

**Speakers:**

– Patrick **Doelle**, European Commission, DG Home, Italy
"The implementation of the 2015 European Agenda on Migration: progress and challenges"

– Andrea Rachele **Fiore**, Puglia Regional Administration, Italy
"Puglia Region policies and initiatives for the reception and integration of immigrants"

– Gianfranco **Gadaleta**, Puglia Regional Administration, Italy
"Territorial cooperation and migration"

– Prof Philip L. **Martin**, University of California, Davis, USA
"Migration policy and labour recruitment"
*Please note this session is by invitation only*

17:30 - 19:15

## Day Two 19 June 2019 - 17:00-18:45

**Aula XX - R. Fonte, 2nd Floor**
**8H. Migration Studies on Display**

Chair    *Mehari Fisseha, Regent's University London, UK*

551    Examining linguistic inequality in AVR repatriation programmes
Katy Brickley

326    Assimilationism, Multiculturalism, Colorblindness, Polyculturalism, and Intergroup Bias in the Russian Context
Dmitrii Dubrov, Dmitriy Grigoryev, Tomas Jurcik, Anastasia Batkhina

372 Migrant Labour in Kerala: Unveiling the Dynamics of In-migration of the Inter-state Migrant Workers
Arun Perumbilavil Anand

334 Exclusion of Muslims in East and West. A comparative analysis of anti-Muslim attitudes in France, Norway, Poland and Czech Republic
Zan Strabac

514 Southern migrant women resistances in the Metropoles: to be a Brazilian woman in Potugal/ to be a Colombian woman in Spain
Andrea Souto Garcia

126 Alternative Asylum and Refugee Narratives: Deconstructing the role of NGO's and Civil Society within an Evolving European Public Sphere
Muhamed Shiwan Amin

68 Diversity in the Classrooms: Assessing the Educational Needs of the Migrant Children in Kerala, India
Arun Perumbilavil Anand

## Day Two 19 June 2019 - 17:00-18:45

**Aula XXI - Don Tonino Bello, 2nd Floor**
**8J. Migration Studies on Display**

Chair   *Mehari Fisseha, Regent's University London, UK*

952 New migrations in small Central European countries
Eva Rievajova

951 Understanding immigrant integration in a post-transition country
Andrej Privara

36 Harnessing Labor Migration Dividends in Africa through Continental Frameworks
Lupwana Jean Jacques Kandala

966 Macroeconomic impact of population change and mobility in Slovakia
Magdaléna Přívarová

699 How to be a [little] fish in a large pond and survive: ethnic diasporas as political entrepreneurs in Canada
Klavdia Tatar

42 Language instruction and inequalities in the integration process of young adult Syrian refugees in Turkey
Maissam Nimer

**Aula XXII -G. Di Vagno, 2nd Floor**
**8K. Migration Studies on Display**

Chair   *Mehari Fisseha, Regent's University London, UK*

769 The Discussion about Turkish Integration Failure in Germany; in the Context of Religion and Identity Conceptions
Dhanimol M. M.

182 Rethinking Refugee Protection in Post-Global Compact on Migration
Nafees Ahmad

335 Kurdish and Palestinian Diasporic Activism in Sweden
Fanny Christou

78 Civil Participation of Immigrants: Syrian Immigrant Sample
Yeliz Polat & Sueda Gurses

79 Immigration And Organization In Turkey:  Turkish Direcdorate Genaral of Migration Management
Yeliz Polat & Ayse Yilmaz

## Day Two 19 June 2019 - 17:00-18:45

**Aula XXIII - T. e V.Fiore, 2nd Floor**
**8L. Migration Studies on Display**

Chair  *Mehari Fisseha, Regent's University London, UK*
460  Conceptualising the Migration-Development Nexus: A Transnational Relational Approach
Xuchun Liu
464  Emigration on the Go: Toward a Theory of Transnational Spontaneity
Abdi Kusow
513  An approach to the analysis of the interrelation between migration and social capital in transnational social space
Laura Suárez-Grimalt & Màrius Domínguez i Amorós
167  From Protection to Rehabilitation- Urban As a New Tool For New Refugee Regime: The Field Study of Afyonkarahisar
Atahan Demirkol
80  The Education Problems of Syrian Immigrant Young People in Turkey: The Example of Fırat University
Yeliz Polat

**Aula V. Starace, 2nd Floor**
**8M. Technology and People on the Move**

Chair  *Emre Eren Korkmaz, University of Oxford, United Kingdom*
259  Transformative Technologies and "Empowerment" of Refugees
Emre Eren Korkmaz
345  The role of smartphones in the migration process: Achieving belonging via information and communication technologies
Marco Marinucci, Luca Pancani, Paolo Riva, Davide Mazzoni
178  The impact of communication technologies on migration intentions
Silvia Migali, Sona Kalantaryan, Marco Schipioni, Sara Grubanov Boskovic
27  Refugees help Refugees [D]
Adam Labaran

## END OF DAY TWO

# Day Three 20 June 2019 - 08:45-10:30

### Aula II - G. Giugni – Ground Floor
### 9A. Remittances

Chair   Liliana Meza Gonzalez, National Institute of Statistics and Geography, Mexico

378   Unfolding Perceptions on Remittance and Development: Experiences and Policy Responses from the State of Kerala
Parvathy Devi K.

781   From Codevelopment to "Triple Win Policies". How Migrant NGOs are Facing the Changes and Continuities of the French Political Pardigm towards Migration and Development
Claire Vincent-Mory

557   Remittance Behavior of Migrant Citizens
Nonna Kushnirovich

906   Remittances and labour supply revisited: New evidence from the Macedonian behavioural tax and benefit microsimulation model
Marjan Petreski

585   Transnational motorways. Trade with secondhand cars in an emigration country
Anatolie Cosciug

### Aula III - G. Di Vittorio – Ground Floor
### 9B. Refugees and Barriers

Chair   Nicole Dubus, San Jose State University, USA

115   Refugees from five countries report on the quality and helpfulness of services provided during resettlement
Nicole Dubus

536   What Do They Say to Asylum Seekers? The Admission Process in Belgian Asylum Seeker Centres
Amandine Van Neste-Gottignies

553   "We know we will face challenges" - Narratives of Syrian Forced Migrants: Topicalised Themes and Navigated Identities
Kinan Noah & Eva-Maria Graf

72   Employment Barriers for Refugees in the Hosting Countries
Cevdet Acu

88   The Legal Implication of the Exclusion Clause on Criminal Refugees
Olawale Iskil Lawal

### Aula IV - P. Labriola, Ground Floor
### 9C. Return Migration

Chair   Thais Garcia Pereiro, Catholic University "Our Lady of Good Counsel", Albania

46   "This is my name": Young Israelis of Ethiopian background and their return to Amharic (heritage) names
Liat Yakhnich & Sophie D. Walsh

264   "I'm in a grey zone": A Narrative Analysis of Return Migration and Ethnic Identity
Irasema Mora-Pablo & Martha Lengeling

928   "Returning home?": The politics and practices of return and the migration-development-nexus
Andrea Frieda Schmelz

209   The role of International Partners in return migration policy-making and implementation: the case of the German Cooperation for Development (GIZ) in Cameroon - Charles Simplice Mbatsogo Mebo

# Day Three 20 June 2019 - 08:45-10:30

### Aula XVI - S. Fumarulo, 2nd Floor
### 9D. Portrayals and Borders

| | |
|---|---|
| Chair | Inci Aksu Kargin, Usak University, Turkey |
| 192 | The Portrayal of Syrian Refugees in Popular Humor Magazines in Turkey<br>İnci Aksu Kargın |
| 39 | Middle Eastern Refugees and Media in Canada<br>Amira Halperin |
| 37 | Press Coverage of the Migration: A Content Analysis of Bosnia and Herzegovina<br>Serap Fiso |
| 87 | Re-mapping the Border of the State & the Academy from Within & Without: The Queer Refugee Camp & the Migratory Text as a Space for Alterity & a Site of Critique [D]<br>Chris Campanioni |
| 401 | Crossing the Borders of the Deficit Gaze? Organizing Subjectivation of "The Refugee" in political and academic Discourses<br>Sepideh Abedi Farizani |

### Aula XVII - S. Bianco, 2nd Floor
### 9E. Family and Migration

| | |
|---|---|
| Chair | Gul Ince Beqo, Catholic University of Milan, Italy |
| 581 | Intergenerational comparison of values of Russians and Georgians in Georgia<br>Ekaterina Bushina, Victoria Galyapina, Nadezhda Lebedeva |
| 768 | A Study of Childrearing Practices among Muslim Immigrant Women Living in Canada<br>Wei-Wei Da |
| 919 | Settled Southern-African migrants parenting experience in destination countries: A Qualitative Synthesis<br>Ruvimbo Machaka |
| 696 | The Right to Family Life and Care: Migrant Domestic Work in the European Union<br>Edit Frenyo |
| 739 | Volunteering and Social Integration of Ex-Soviet Immigrants in Israel<br>Natalia Khvorostianov & Larissa Remennick |

### Aula XVIII - N. Calipari, 2nd Floor
### 9F. Din ve Göç [TURKISH]

| | |
|---|---|
| Chair | Yakup Çoştu, Hitit University, Turkey |
| 252 | Almanya'da İslamofobi'nin Artışında Neo-Selefi Akımların Rolü ve Müslüman Göçmenlere Etkileri<br>Mehmet Akif Ceyhan |
| 241 | Securitization Theory and European Muslim Communities<br>Feyza Ceyhan Çoştu & Yakup Çoştu |
| 500 | The Bounty of the East to Andalusia: Andalusian Music A Review on Immigrant Musicians and Singers<br>Sema Dinç |
| 426 | Migration and Socio Cultural Interaction<br>Sıddık Korkmaz |

# Day Three 20 June 2019 - 08:45-10:30

**Aula XIX - L. Ferrari Bravo, 2nd Floor**
**9G. From Move to Settlement**
Chair *Camila Escudero, Universidade Metodista de São Paulo, Brazil*
445 Refugee-based reasons in refugee resettlement
Annamari Vitikainen
577 The Sri Lankan Tamil Migrants: Shifting ideologies from asylum seekers
to a prominent diaspora community
Tulika Gaur
260 Job Thieves: Migrations and the Labor Market in South Africa
Yolanda Lohelo Emedi
40 Unskilled labour migration, control and compliance: Private agencies in
South Korea and Thailand
Steve Kwok-Leung Chan
323 Migrants or Refugees?: An analysis of the recent wave of population
movements from Venezuela to Brazil
Gisela P. Zapata & Sulma Marcela Cuervo

**Aula XX - R. Fonte, 2nd Floor**
**9H. Discourse and Representation in Media**
Chair *Süreyya Sönmez Efe, University of Lincoln, United Kingdom*
775 Media framing of Syrian refugees in Europe: A comparative analysis of
news reporting in terms of peace journalism
Aynur Sarısakaloğlu
387 Media, Politics and Power in the Representation of the Refugee and
Migration Crisis in Europe
Evangelia Tomara
165 Immigration in Belgian television news: few differences between public
and commercial broadcasters?
Valériane Mistiaen
174 Three years of the Refugee crisis's representation: the public opinion on
Facebook
Dario Lucchesi

**Aula XXI - Don Tonino Bello, 2nd Floor**
**9J. Management of Migration and Integration Process: Legal**
**migrants and their integration in Italy [Master Programme Students]**
**[D]**
Chair *Daniele Petrosino, University of Bari, Italy*
989 Integration by working: job placement into the Italian reception system for
asylum seekers and refugees
Marta Chiffi
987 Asylum law and integration policies: The "Bari Città Aperta" Sprar Project
Giulia Fasiello
986 Refugees Law in Italy: authority and procedure
Andrea Mongelli
988 Exposure to infectious agents among Italian State Policemen engaged in
activities to cross-border immigrants: risk perception versus risk
assessment
Carmela Montrone
985 Italy-Nigeria Round Trip Ticket: Projects of development and integration
Antonhy Nwiboko

# Day Three 20 June 2019 - 08:45-10:30

**Aula XXII -G. Di Vagno, 2nd Floor**
**9K. Göç ve Küreselleşme [TURKISH] [D]**
Chair  Fatih Turan Yaman, İstanbul Rumeli University, Turkey
981  Küresel Yönetişim ve Göç
Fatih Turan Yaman, Emrullah Kocabaş
982  Küresel İklim Değişikliğinin Küresel Göç Üzerinde Olabilecek Muhtemel Etkisi - Fethi Furkan Elbir, Burak Çamur
983  SSCB ve Sonrası Orta Asya'da Göç Hareketleri
Halil Günay
984  Uluslararası Göçün Küreselleşmeye Etkileri
Semanur Polat

**Aula XXIII - T. e V.Fiore, 2nd Floor**
**9L. Nuevas tendencias migratorias en América Latina y el Caribe: Una mirada a los conflictos y dinámicas de salida, retorno, re - asentamiento, derechos humanos y políticas públicas [SPANISH]**
Chair  Maria Rocio Bedoya Bedoya, Universidad de Antioquia, Colombia
936  Migración venezolana hacia Colombia: respuestas del Estado a la llegada masiva entre 2015 y 2018
Maria Rocio Bedoya Bedoya
137  Ethnosurvey on Recent Immigration. An experience to address hard-to-reach immigrant groups in a context of South-South migration
Victoria Prieto & Clara Márquez
927  Las caravanas migrantes por México: ¿estrategias de movilidad en el siglo XXI?
Cristina Gómez Johnson
937  ¿De receptores a emisores y al revés? Retos sociojurídicos de la migración venezolana reciente en la región latinoamericana
Luciana Gandini, Fernando Lozano-Ascencio & Victoria Prieto
939  Itinerarios del sentido de lugar: el quedarse, el salir y el volver en contextos de movilidad forzada. Colombia y México 2000- 2017
Laura Cartagena Benítez
938  Desplazados Internos Por La violencia Y Ciudad: La espacialización Urbana Como Aplazamiento Indefinido Del Goce Efectivo De Los Derechos
Clara Atehortúa-Arredondo

**Aula  - V. Starace, 2nd Floor**
**9M. Arts and Migration**
Chair  Vildan Mahmutoglu, Galatasaray University, Turkey
901  Image as trajectory in the art installation Outside in: exile at home
Annabel Castro
64  Body modification, Assimilation, and Integration in the Story of Joseph
Eric M. Trinka
650  Migration and Literature
Angela Zagarella

**10:30-10:45 – Break**

## Day Three 20 June 2019 - 10:45-12:30

### Plenary Session III: Reflecting on World Refugee Day
**Aula Aldo Moro, Ground Floor**
Chair: Isabella **Piracci**, Avvocatura Generale dello Stato, Rome, Italy
**Speakers:**
– Antonio **Di Muro**, UNHCR representative, Italy
– Jeffrey H. **Cohen**, Ohio State University, USA
   "A house of mirrors: refugees and reflection"
– Ferruccio **Pastore**, FIERI – Forum of International and European Research on Immigration, Italy
   "The "mixed migration" dilemma and the roots of Europe's migration and asylum governance crisis"
– Giuseppe **Albenzio**, Vice Avvocato Generale dello Stato, Italy
   "The recent reform of the law on immigration"

*(left margin: 10:45-12:30)*

### Plenary Session IV: Technology and People on the Move
**Aula II - G. Giugni – Ground Floor**
Chair: Emre Eren **Korkmaz**, University of Oxford, United Kingdom
**Speakers:**
– Marie **Godin** & Giorgia **Dona**, Forced migrants in transit: mobile technologies and techno-borderscapes
– Aiden **Slavin**, ID2020, Digital identity for refugees: is blockchain technology a panacea?
– Albert Ali **Salah**, Boğaziçi, Using mobile phone data for the analysis of refugee movements: The Data for Refugees Challenge for Syrian Refugees in Turkey
– Zeynep Sıla **Sönmez**, UNHCR, Meet the future: Digital identity initiatives as humanitarian solutions

*(left margin: 10:45-12:30)*

**12:30-13:30 – Lunch**

### Graduation Ceremony for Masters in Management of Migration and Integration Process
**Aula Magna – Palazzo Ateneo, Ground Floor**
*Convener: Prof Michela C. **Pellicani**, University of Bari, Italy*

*(left margin: 13:30)*

## Day Three 20 June 2019 - 13:30-15:00

**Aula Aldo Moro, Ground Floor**
**10X. SPECIAL CONVERSATION ON CINEMA**
214    *Can Cinema Tell about Migration? - Annalisa **Morticelli***

**Aula II - G. Giugni – Ground Floor**
**10A. Migration and Integration**
Chair    *Karolina Nikielska-Sekula, University College of Southeast Norway*
270    Integration challenges: case of Italy
       Mariann Dömös
135    Contemporary Immigrants in South Africa: Perceptions, Treatment, and Integration
       Terry-Ann Jones
235    Armed conflicts and the dynamics of refugee migrations: Conceptualizing Syrian refugee migrations
       Marko Valenta

# Day Three 20 June 2019 - 13:30-15:00

**Aula III - G. Di Vittorio – Ground Floor**
**10B. Skilled Migration**

Chair  *Nirmala Devi Arunasalam, University of Plymouth, UK*

449  Teachers without Borders: The transnational migration of South African teachers
Sadhana Manik

944  1st year student nurses' views of Link Lecturers versus Clinical Professors
Nirmala Devi Arunasalam, Thayer McGahee, Betty Abraham-Settles

558  STEM track enrolment among second-generation immigrant students from high-skilled parental background
Svetlana Chachashvili Bolotin & Sabina Lissitsa

598  The effect of the Diaspora-Homeland relationship on the intellectual reverse brain drain: A research on Turkish scientists in Germany
Atakan Durmaz & Adem Kalça

**Aula IV -  P. Labriola, Ground Floor**
**10C. Migration and Migrants in Media**

Chair  *Vildan Mahmutoglu, Galatasaray University, Turkey*

3  Cultural Proximity or Cultural Distance?: Selecting Media Content Among Turkish Diasporic Audiences in Germany
Miriam Berg

948  A Critical Analysis of Public-based Discourse about International Migration and Immigrants
M. Murat Yüceşahin & İrfan Tatlı

676  Media representation of Romanian migrants in the Italian media: a comparative study
Paula Catalina Meirosu

542  The effects of cognitive mobilization, media salience, and party identity on immigration attitudes
Maija Ozola, Sherief Emam, Jens Wolling

**Aula XVI -  S. Fumarulo, 2nd Floor**
**10D. Onward Migration in a Changing Europe**

Chair  *Nicola Montagna, Middlesex University, United Kingdom*

704  Secondary movements in the European Union: tracing the evidence
Simon McMahon

403  Multiple migrations, Networks, and Transnational Ties: The Case of Italian Bangladeshi Onward Migrants in Europe
Mohammad Morad & Devi Sacchetto

391  Roots and routes of Brazilian migration to and within the European Union
Pedro Gois & Jose Marques

144  Brexit as a trigger and an inhibitor of onward and return migration
Djordje Sredanovic

# Day Three 20 June 2019 - 13:30-15:00

### Aula XVII - S. Bianco, 2nd Floor
### 10E. Return Migration
Chair    Merita Zulfiu Alili, South East European University, N. Macedonia
541    Communication Regarding Voluntary Return Programmes in Belgian Asylum Seekers Reception Centres: From Mediated Forms of Communication to Face-to-face Interactions
Amandine Van Neste-Gottignies & Valériane Mistiaen
540    The Gender Dimension of Return Migration: Turkish Skilled Returnees from Germany and the US
Meltem Yilmaz Sener
386    Portuguese migration to the South: a return to the past in a postcolonial world?
Jose Carlos Marques & Pedro Góis
184    "Migrants" eternal returns: Greek migrants' return visits from West Germany to Greece as a prelude to their return migration"
Maria Adamopoulou

### Aula XVIII - N. Calipari, 2nd Floor
### 10F. Migration and Policy
Chair    Ülkü Sezgi Sözen, Universität Hamburg, Germany
114    Strong cases, negative decisions. Asylum assessment as border control
Erna Bodström
498    The Ethical Consequences of Criminalizing Solidarity with Refugees and Asylum Seekers for Western Democracies
Melina Duarte
717    Restrictive Asylum Policies and Reflections in the Labour Market: The cases of Italy and Turkey
Secil E. Ertorer and Anita Butera
554    The challenge of today's Common European Asylum Area : a re-appropriation of the balance between claims of national security and fundamental rights
Roila Mavrouli

### Aula XIX - L. Ferrari Bravo, 2nd Floor
### 10G. Remittances and Migration
Chair    Pooja Batra, Indian Institute of Management (IIM), Indore, India
903    Squandering Remittances Income in Conspicuous Consumption: The Case of Uzbekistan
Jakhongir Kakhkharov & Muzaffar Ahunov
307    Internal versus International Migration and the consumption well-being of Left behind Households in Kerala
Ajay Sharma & Pooja Batra
437    The effect of remittance on household spending behaviour in East Lombok, Indonesia
Nurlia Listiani & Temesgen Kifle
8    Gulf Migration and The Flow of Social Remittances: "Arabization" among Mappila Muslims of Malabar
Mohamed Ameen Arimbra

# Day Three 20 June 2019 - 13:30-15:00

## Aula XX - R. Fonte, 2nd Floor
### 10H. Gender and Migration
Chair  *H. Yaprak Civelek, Istanbul Arel University, Turkey*
309  New Understanding of Women's Historical Migration
Tomoko Tsuchiya
684  The paradox of empowerment among Syrian refugees in Lebanon
Irene Tuzi
548  Straints of the Syrian Refugee Crisis on the Gender Gap of Lebanon's labour market [D]
Laura El Chemali
709  Motherhood of Nigerian and Romanian women in sexual exploitation
Rafaela Pascoal

## Aula XXI - Don Tonino Bello, 2nd Floor
### 10J. Göç ve Edebiyat [TURKISH]
Chair  *Yakup Çelik, Istanbul Kultur University, Turkey*
1001  Halikarnas Balıkçısı'nın Romanlarının Göç Kavramı Çevresinde Değerlendirilmesi
Yakup Çelik
1003  Kemal Tahir'in *Bir Mülkiyet Kalesi* Romanında Göçün Sürekliliği ve Dönüşümü
Güler Uğur
1004  Hasan Ali Toptaş'ın Kuşlar Yasına Gider Romanında Yolculuk
Berna Civalıoğlu

## Aula XXII -G. Di Vagno, 2nd Floor
### 10K. Family and Migration
Chair  *Gul Ince Beqo, Catholic University of Milan, Italy*
347  Watching Them Grow': Video Chatting as a Household Practice Among Migrant Children and Remote Relatives
Ipek Demirsu
397  Family Change Model and Attitudes towards the Filial Obligation Norms Among Immigrant and Host Groups
Nagihan Taşdemir
457  Labour Migration and Its Impact on Rural livelihoods in Tajikistan, 1997-2017
Neha Tewari
471  Chinese filial piety across migration, a comparison between Wenzhouese migrants and stayers
Nicoletta Galisai

# Day Three 20 June 2019 - 13:30-15:00

**Aula XXIII - T. e V.Fiore, 2nd Floor**
**10L. Identity and Representation**
Chair  Pinar Yazgan, Sakarya University, Turkey
624  Hate speech against Muslims on social media in Italy. Evidence from the European project Hatemeter
Elisa Martini & Gabriele Baratto
905  The Dynamics of Refugees' Dual- Identity along Ethio-South Sudan Border: Challenges, Prospects and Policy Implications
Moti Gutema
352  The Mechanism and Factor Analysis of Anti-Immigration Sentiment Dissemination in Social Media: A Comparative Study in U.S., U.K., and Japan
Keisuke Idemitsu
53  Refugee Children and Adolescents in the Italian Media. Reflections on narrative and representation [VIDEO]
Valentina Baú

**Aula - V. Starace, 2nd Floor**
**10M. Göç ve Aidiyet [TURKISH] [D]**
Chair  Süleyman Özmen, Istanbul Rumeli University, Turkey
1009  Göçmenleştirme Politikalarının Devletlerin İstikrarsızlaştırılmasına Etkileri
H. Murat Lehimler, Süleyman Özmen
1010  Türkiye Avrupa Göç Sistemi Misafir İşçilikten Ulus Ötesi Kimliklere
Hatice Sırtıkara
1011  Aidiyet Duygusunun ve Göç Sorunsalı Üzerine Etkileri
Mert Bilecen

**Aula V. Buono, 6th Floor**
Special Session: Percorsi di accoglienza e integrazione dei cittadini stranieri [ITALIAN]
Organised in collaboration with Caritas – Diocesi Bari Bitonto
Chair: Don Vito **Piccinonna**, Director of Caritas - Diocesi Bari Bitonto, Italy
**Speakers:**
– Emanuele **Abbatantuono**, Fondazione Opera Santi Medici Cosma e Damiano Bitonto ONLUS
"Accoglienza tra solidarietà e sussidiarietà"
– Vito **Mariella**, Ass. Micaela onlus
"Tratta e migrazioni"
– Alessandra **Pupillo**, Centro di accoglienza "don Vito Diana" - coop. Soc. Equal Time
"Senza dimora italiani e stranieri"
– Valeria **Sasanelli**, Casa Ain Karem - coop. Soc. Mi stai a cuore
"Accoglienza e inserimenti sociali"

13:30-15:00

**15:00-15:15 BREAK**

# Day Three 20 June 2019 - 15:15-16:45

### Aula II - G. Giugni – Ground Floor
### 11A. Displacement Challenges

Chair   *Simeon Magliveras, King Fahd University of Petroleum and Minerals, Saudi Arabia*

73   Humanitarian crisis of Central Americans migrants in transit through Mexico: forced migration and accompaniment process
Jorge Morales Cardiel

393   The Syrian Refugee Crisis and the Refugee Rentier State in the Middle East
Gerasimos Tsourapas

597   "From that moment onwards, we became the Cerberus inside the camp": The Role of Local Political Leadership in Refugee Reception
Tihomir Sabchev

567   Immigration policy and party politics in Italy: beyond left-right polarisation?
Audrey Lumley-Sapanski & Senyo (Noah) Dotsey

### Aula III - G. Di Vittorio – Ground Floor
### 11B. Migration and Integration

Chair   *Niamh Dillon, University of Limerick, Ireland*

941   Migration Nation: Barriers to integration for asylum-seekers in Ireland
Niamh Dillon

455   The 28-DAY Program:  a refugee integration support program in Liverpool, England
Chien-Yi Chu

655   Migrant integration. A way to reverse the demographic decline in Europe
Razvan Dacian Carciumaru

771   Cash Based Interventions as a Refugee Integration Facilitator
Selman Salim Kesgin

### Aula IV -  P. Labriola, Ground Floor
### 11C. Arts and Migration

Chair   *Persefoni Myrtsou, Humboldt University Berlin, Germany*

507   Migrant Artists: Spatialised Self-images and Practices of (Un)Belonging
Persefoni Myrtsou

524   Spaces of Musical Production as Cultural Borderlands
Carolin Mueller

762   Graphic Narratives: Representations of Refugeehood in the Form of Illustration
Pauline Blanchet

1201   Writing, music, migration: from Hanif Kureishi to David Bowie
Pierpaolo Martino

### Aula XVI - S. Fumarulo, 2nd Floor
### 11D. Migration and Education

Chair   *Silvia Juarez-Marazzo, Chances for Children-NY, USA*

923   How Children's Picture Books may support Latino American asylum seeking mothers restore their children's hope and sense of trust?
Silvia Juarez-Marazzo

493   Russia's Place In The World Through The Eyes Of Foreign Students
Daria Borisovna Kazarinova, Arussyak Hauvhanissyan, Vasilya Taisheva

539   Between diaspora and state: The case of an informal immigrant school in Halkalı
Cigdem Billur Ada

705   Family-School Relations and Trust in an Intercultural Context
Mina Prokic

# Day Three 20 June 2019 - 15:15-16:45

### Aula XVII - S. Bianco, 2nd Floor
### 11E. Göç ve Kültür [TURKISH] [D]
Chair   Selin Özmen, Tekirdağ Namık Kemal University, Turkey
1005    Göç ve Kültürel Etkileşim: M.Ö. I. Binde Bezemeli Lydia Seramiğinde Ionia Etkisi - Selin Özmen
1006    Gelenek - Görenek Karşılaştırması ve Göç - Selin Özdemir
1007    Beyin Göçü ve Türkiye - Vesile Topsakal
1008    Why are we so resistant about learning immigrants' languages?
        Tuba Tekin

### Aula XVIII - N. Calipari, 2nd Floor
### 11F. Transformations with Migration
Chair   Carlo Alberto Anzuini, University of Bari, Italy
780     The local turn in integration policies for migrants and refugees in Latin America
        Joao Guilherme Casagrande Martinelli Lima Granja Xavier da Silva
574     Vacaciones en Paz: Sahrawi refugees between political activism and Spanish solidarity
        Rita Reis
348     The Transformative Power of Work with Refugees and Asylum Seekers for Practitioners of NGOs
        Lina Grudulaita
379     Chronic illness management in the face of adversity: experiences from asylum seekers with diabetes in the Belgian reception centers
        Wanda Van Hemelrijck

### Aula XIX - F. Bravo, 2nd Floor
### 11G. Migration and Integration
Chair   Gul Ince Beqo, Catholic University of Milan, Italy
197     Art and Integration - A Civic Collaboration
        Jeannie Simms
254     "Transfer of Ritual" and Struggle for Recognition Among Ethiopian Immigrants
        Rachel Sharaby
925     Diaspora Integration through consular services: the case of Mexican Consulates in Sacramento and San Francisco, California
        Karla A. Valenzuela Moreno
358     Homeland vs Hostland: A Comparative Study on the Relation of Post-Migrant Populations with Turkey Background along Right Populisms in Germany and Italy
        Gul Ince Beqo, Asli Telli Aydemir, Ali Ekber Dogan

### Aula XX - R. Fonte, 2nd Floor
### 11H. Migration Law and Policy
Chair   Pinar Yazgan, Sakarya University, Turkey
489     Autonomy through law? The ambivalent role of law in the German reception facilities - Anne-Marlen Engler
755     History, Race & the Politics of Belonging | How Politicized Responses to Migrants and Refugees in Southern Europe Are Undermining Economic Development, Social Integration & International Standards in Humanitarian Law - Abigail J Blue
322     Migration policy in Greece: between a transit and an integration society
        Evangelia Kasimati & Roy Panagiotopoulou
929     The State as a Birthplace and Main Cause Behind Massive Migration and Refugee Outflows [VIDEO]
        Karaman Mamand

# Day Three 20 June 2019 - 15:15-16:45

### Aula XXI - Don Tonino Bello, 2nd Floor
### 11J. Family and Migration
Chair   Monica Ibáñez-Angulo, University of Burgos, Spain
103   A report on transnational marriages and migrant workers at the China-Vietnam and China-Myanmar border areas
Xueqing Zhao, Shenghui Yang, Yimeng Zhao
229   Unpacking The Meaning of Marriage Among Transnational Couples: A Multiadic Analysis
Elizabeth Marfel Fortes & Gagni Edith Liane PenaAlampay
312   Immigrant Women Intermarriage Premium: Italy 2006-2012
Adda Carla Justiniano
399   Gender and race among young adults of Nigerian descent in Ireland
Inga Wójcik


### Aula XXII -G. Di Vagno, 2nd Floor
### 11K. Work, Employment, Markets [D]
Chair   Liudmila Konstants, American University in Central Asia, Kyrgyzstan
331   Precarious Employment of International Migrant Workers: The State and Ways to Reduce
Igor Alexeevich Shichkin
332   The Pull-Push Factors of Migration: Views and Experiences of the Pakistani Migrant Workers in Saudi Arabia
Muhammad Saeed
646   The effect of the Experience of being in poverty on Individual Psychological Characteristics
Maria Viktorovna Efremova, Olga Vladimirovna Poluektova, Seger Breugelmans
774   Precarious Mobility in India: Ethnographic evidence from the Mising Community of Assam
Bipul Pegu
227   Migration Diversity and House Prices - Evidence from Sweden
Adam Alexander Tyrcha


### Aula XXIII - T. e V.Fiore, 2nd Floor
### 11L. Göç ve Türkiye [TURKISH]
Chair   Esin Hamdi Dinçer, Artvin Çoruh University, Turkey
1017   Suriyeli Emeğine Yönelik İşçi Örgütlerinin Deneyimleri: Çatışma ve Uyum Ekseninde Bir İnceleme
Fethiye Tilbe
748   Türkiye'deki Suriyeliler ve Agamben'in Siyaset Felsefesi
Esin Hamdi Dinçer
1018   Ulusal Güvenlik Perspektifinde Suriyeli Göçmenlere Yönelik Sosyal Medya Paylaşımları
Cebrail Aydın
1019   The Structure and Current Problems of Syrian Immigrants
Gamze Karagöz
1020   1950'den günümüze Türkiye'de İç Göç ve Düzensiz Kentleşme
H.İbrahim Karakuş, Rüstem İnan

## Day Three 20 June 2019 - 17:00-18:45

**17:00 - 18:45**

### Roundtable IV: Migration Policy in Italy and Beyond
In collaboration with and support from
**Aula Aldo Moro, Ground floor**

Chair: Michela C. **Pellicani**, University of Bari Aldo Moro, Italy
**Speakers:**
– Massimo **Bontempi**, Central Director of the Immigration and Border Police, Home Affairs Ministry, Italy
– Luigi Maria **Vignali**, General Director for the Italians abroad and the Migration Policies, Foreign Affairs Ministry, Italy
– Fabrice **Leggeri** (tbc), Executive Director, FRONTEX
– Jeffrey H. **Cohen**, Ohio State University, USA
– Ibrahim **Sirkeci**, Regent's University London, UK

## Day Three 20 June 2019 - 18:45-19:15: Closing Remarks
[ **Aula Aldo Moro, Ground Floor** ]

## Day Three 20 June 2019 - 20:00-22:30: GALA DINNER
**Circolo Unione - Teatro Petruzzelli**, Corso Cavour, 12, 70122 Bari, Italy.
It is about 10 minutes walking distance from the conference venue.
Tel: + 39 (080) 521 1249 | https://www.facebook.com/pages/category/Arts---Entertainment/Circolo-Unione-Teatro-Petruzzelli-392710240812854/

# www.migrationconference.net

# #tmc19bari

# Name Index